Organic Chemistry II

by Frank Pellegrini, Ph.D.

IDG Books Worldwide, Inc.
An International Data Group Company
Foster City, CA ♦ Chicago, IL ♦ Indianapolis, IN ♦ New York, NY

D1166818

About the Author

Dr. Frank Pellegrini has been a professor for over 35 years. His field of specialization is heterocyclic organic chemistry. He is the recipient of the State University of New York Chancellor's Award for Excellence in Teaching. Currently he holds the rank of Professor of Organic Chemistry and Department Chair at SUNY Farmingdale. Other interests include photography, travel, and volunteer fire fighting.

Publisher's Acknowledgments

Editorial

Project Editor: Joan Friedman

Acquisitions Editor: Kris Fulkerson

Copy Editor: Billie A. Williams

Technical Editor: Benjamin Burlingham

Editorial Assistant: Laura Jefferson

Production

Proofreader: Christine Pingleton

IDG Books Indianapolis Production Department

CliffsQuickReview Organic Chemistry II

Published by
IDG Books Worldwide, Inc.
An International Data Group Company
919 E. Hillsdale Blvd.
Suite 400
Foster City, CA 94404

www.idgbooks.com (IDG Books Worldwide Web site)
www.cliffsnotes.com (CliffsNotes Web site)

Library of Congress Catalog Control Number: 00-103367

ISBN: 0-7645-8616-5

Printed in the United States of America

10 9 8 7 6 5 4 3 2 1

1O/QV/QX/QQ/IN

Distributed in the United States by IDG Books Worldwide, Inc.

Distributed in Canada by CDG Books Canada Inc. for Canada; by Transworld Publishers Limited in the United Kingdom; by IDG Norge Books for Norway; by IDG Sweden Books for Sweden; by IDG Books Australia Publishing Corporation Pty. Ltd. for Australia and New Zealand; by TransQuest Publishers Pte Ltd. for Singapore, Malaysia, Thailand, Indonesia, and Hong Kong; by Gotop Information Inc. for Taiwan; by ICG Muse, Inc. for Japan; by Intersoft for South Africa; by Eyrolles for France; by International Thomson Publishing for Germany, Austria and Switzerland; by Distribuidora Cuspide for Argentina; by LR International for Brazil; by Galileo Libros for Chile; by Ediciones ZETA S.C.R. Ltda. for Peru; by WS Computer Publishing Corporation, Inc., for the Philippines; by Contemporanea de Ediciones for Venezuela; by Express Computer Distributors for the Caribbean and West Indies; by Micronesia Media Distributor, Inc. for Micronesia; by Chips Computadoras S.A. de C.V. for Mexico; by Editorial Norma de Panama S.A. for Panama; by American Bookshops for Finland.

For general information on IDG Books Worldwide's books in the U.S., please call our Consumer Customer Service department at **800-762-2974.** For reseller information, including discounts and premium sales, please call our Reseller Customer Service department at **800-434-3422.**

For information on where to purchase IDG Books Worldwide's books outside the U.S., please contact our International Sales department at **317-596-5530** or fax **317-572-4002.**

For consumer information on foreign language translations, please contact our Customer Service department at **1-800-434-3422**, fax 317-572-4002, or e-mail rights@idgbooks.com.

For information on licensing foreign or domestic rights, please phone **+1-650-653-7098.**

For sales inquiries and special prices for bulk quantities, please contact our Order Services department at **800-434-3422** or write to the address above.

For information on using IDG Books Worldwide's books in the classroom or for ordering examination copies, please contact our Educational Sales department at **800-434-2086** or fax **317-572-4005.**

For press review copies, author interviews, or other publicity information, please contact our Public Relations department at **650-653-7000** or fax **650-653-7500.**

For authorization to photocopy items for corporate, personal, or educational use, please contact Copyright Clearance Center, 222 Rosewood Drive, Danvers, MA 01923, or fax **978-750-4470.**

is a registered trademark under exclusive license to IDG Books Worldwide, Inc. from International Data Group, Inc.

For the purpose of this review, your knowledge of the following fundamental ideas is assumed.

- Atomic structure
- Lewis structures
- Ionic bonding
- Covalent bonding and electronegativity
- Brønsted-Lowry theory of acids and bases
- Lewis theory of acids and bases
- Mechanisms
- Bond rupture and formation
- Structure of organic molecules
- Structure, properties, and reactions of alkanes, alkenes, alkynes, and cyclohydrocarbons
- Stereochemistry
- Conjugated dienes
- Carbocations, carbanions, free radicals, and carbenes

If you need to review any of these topics, refer to *CliffsQuickReview Organic Chemistry I.*

Introduction

Aromatic compounds are a class of hydrocarbons that possess much greater stability than their conjugated unsaturated system suggests. The simplest example of this class of compounds, benzene, was isolated from illuminating gas by Michael Faraday in 1825. In the years to follow, this compound and homologues were isolated by the distillation of resin gums from balsam trees. Because many of the resin gums had fragrant aromas, these compounds were often called aromatic compounds or aromatic hydrocarbons. In 1845, August Von Hofmann isolated benzene from coal tar. This isolation method remained the chief source of benzene until the 1950s. Today, most benzene is produced from petroleum.

Benzene

In 1834, Eilhardt Mitscherlich conducted vapor density measurements on benzene. Based on data from these experiments, he determined the molecular formula of benzene to be C_6H_6. This formula suggested that the benzene molecule should possess four modes of unsaturation because the saturated alkane with six carbon atoms would have a formula of C_6H_{14}. These unsaturations could exist as double bonds, a ring formation, or a combination of both.

Structure of the benzene molecule

In 1866, August Kekulé used the principles of structural theory to postulate a structure for the benzene molecule. Kekulé based his postulation on the following premises:

- The molecular formula for benzene is C_6H_6.

- All the carbons have four bonds as predicted by structural theory.

- All the hydrogens are equivalent, meaning they are indistinguishable from each other.

Based on these assumptions, Kekulé postulated a structure that had six carbons forming a ring structure. The remaining three modes of unsaturation were the result of three double bonds alternating with three single bonds. This arrangement allowed all the carbon atoms to have four bonds as required by structural theory.

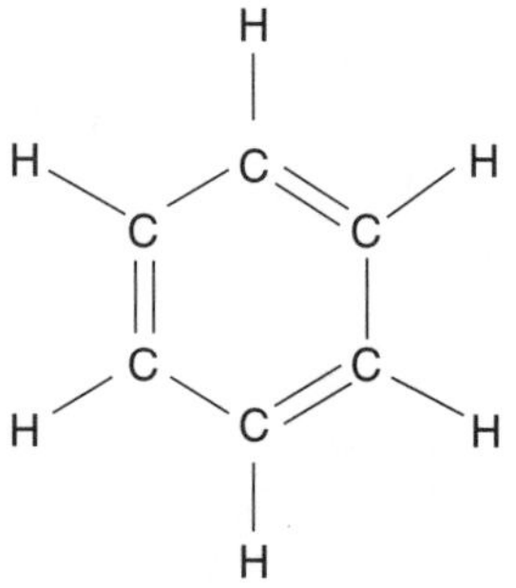

Scientists soon realized that if Kekulé's structure were correct, substituting substituent groups for hydrogens on the 1,2 positions would lead to a different compound than substitution on the 1,6 positions.

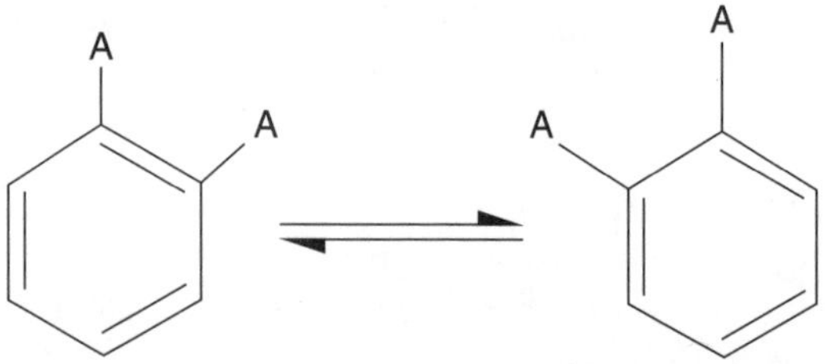

1,2-disubstitution   1,6-disubstitution

Because no such isomers could be produced experimentally, Kekulé was forced to modify his proposed structure. Kekulé theorized that two structures existed that differed only in the location of the double bonds. These two structures rapidly interconverted to each other by bond movement.

Although Kekulé's structure accounted for the modes of unsaturation in benzene, it did not account for benzene's reactivity.

Resonance

Modern instrumental studies confirm earlier experimental data that all the bonds in benzene are of equal length, approximately 1.40 pm. (A **picometer** equals 1×10^{-12} meter.) This bond length falls exactly halfway between the length of a carbon-carbon single bond (1.46 pm) and a carbon-carbon double bond (1.34 pm). In addition, these studies confirm that all bond angles are equal (120°) and that the benzene molecule has a planar (flat) structure.

Modern descriptions of the benzene structure combine resonance theory with molecular orbital theory.

Resonance theory postulates that when more than one structure can be drawn for the same molecule, none of the drawn structures is the correct structure. The true structure is a hybrid of all the drawn structures and is more stable than any of them. The greater the number of structures that can be drawn for a molecule, the more stable the hybrid structure will be. The difference between the calculated energy for a drawn structure and the actual energy of the hybrid structure is called the **resonance energy.** The greater the resonance energy of a compound, the more stable the compound.

The two Kekulé structures that can be drawn for the benzene molecule are actually two resonance structures.

The hybrid of these structures would be drawn as

where the circle represents the movement of the electrons throughout the entire molecule. This delocalization of π **electrons** (electrons found in π molecular orbitals) is also found in conjugated diene systems. Like benzene, the conjugated diene systems show increased stability.

Because of resonance, the benzene molecule is more stable than its 1,3,5-cyclohexatriene structure suggests. This extra stability (36 kcal/mole) is referred to as its **resonance energy**.

Orbital picture of benzene

Because experimental data shows that the benzene molecule is planar, that all carbon atoms bond to three other atoms, and that all bond angles are 120°, the benzene molecule must possess sp^2 hybridization. With sp^2 hybridization, each carbon atom has an unhybridized atomic p orbital associated with it. The overlap of the sp^2 hybrid orbitals would create the σ **bonds** that hold the ring together, while the side-to-side overlap of the atomic p orbitals can occur in both directions, leading to complete delocalization in the π system. This complete delocalization adds great stability to the molecule. Figure 1–1 illustrates this idea.

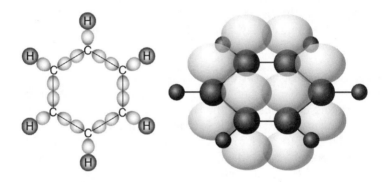

Figure 1–1

Molecular orbital theory predicts that overlapping six atomic p orbitals will lead to the generation of six π molecular orbitals. Three of these π molecular orbitals will be bonding orbitals, while the other three will be antibonding orbitals, as shown in Figure 1–2.

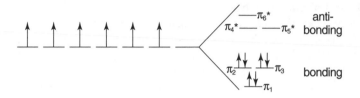

Figure 1–2

The three low-energy orbitals, denoted π_1, π_2, and π_3, are bonding combinations, and the three high-energy orbitals, denoted π_4^*, π_5^*, and π_6^*, are antibonding orbitals. Two of the bonding orbitals (π_2 and π_3) have the same energy, as do the antibonding orbitals π_4 and π_5. Such orbitals are said to be **degenerate.**

Because the electrons are all located in bonding orbitals, the molecule is very stable. Additional stability occurs because all the bonding orbitals are filled and all the π electrons have paired spins. Molecules that possess all these characteristics are said to have a closed bond shell of delocalized π electrons. Molecules such as benzene that possess a closed bond shell of delocalized π electrons are extremely stable and show great resonance energies.

Hückel's Rule

In 1931, Erich Hückel postulated that monocyclic (single ring) planar compounds that contained carbon atoms with unhybridized atomic p orbitals would possess a closed bond shell of delocalized π electrons if the number of π electrons in the molecule fit a value of $4n + 2$ where n equaled any whole number. Because a closed bond shell of π electrons defines an aromatic system, you can use Hückel's Rule to predict the aromaticity of a compound. For example, the benzene molecule, which has 3 π bonds or 6 π electrons, is aromatic.

Number of π electrons $= 4n + 2$

$$6 = 4n + 2$$

$$n = 1$$

However, 1,3,5,7-cyclooctatetraene, which has 4 π bonds or 8 π electrons, is not only nonaromatic but is actually considered *antiaromatic* because it is even less stable than the open-chain hexatriene.

Number of π electrons $= 4n + 2$

$$8 = 4n + 2$$

$$n = 1.5$$

Nomenclature

In IUPAC nomenclature, benzene is designated as a parent name. Other compounds that contain the benzene molecule may be considered as substituted benzenes. In the case of **monosubstitution** (the replacement of a single hydrogen), the prefix of the substituent is added to the name benzene.

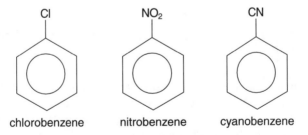

| CI | NO$_2$ | CN |
| chlorobenzene | nitrobenzene | cyanobenzene |

In other cases, the substituent, along with the benzene ring, forms a new parent system.

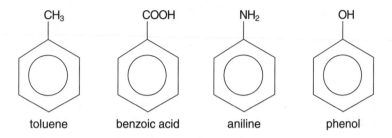

toluene   benzoic acid   aniline   phenol

When a benzene molecule is **disubstituted** (two hydrogens are replaced), two nomenclature methods exist. Either a number system or name system indicates the relative position of one substituent to the other. In the number system, one substituent is given the number one position and the second substituent is assigned the lower possible second number. The number position is given to the atom or group that has the higher priority as determined by the Cahn-Ingold-Prelog nomenclature system rules.

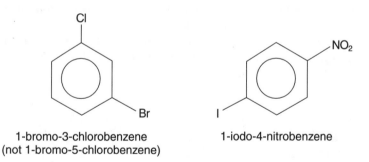

1-bromo-3-chlorobenzene   1-iodo-4-nitrobenzene
(not 1-bromo-5-chlorobenzene)

Notice that in the previous examples, the atom of the higher atomic weight is given the higher priority (Br = 79.1 versus Cl = 15.5, and I = 126.0 versus N = 14.0). These assignments are based on the *priority rules* of Cahn-Ingold-Prelog nomenclature.

In the name system, one carbon atom containing a substituent is considered to be the initial (locator) position. The carbon atom bonded to the other substituent is then located by the number of carbon atoms separating it from the locator position, as shown in Figure 1–3.

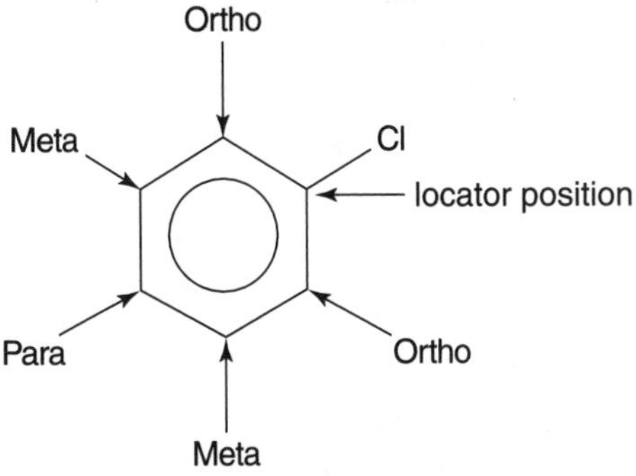

Figure 1–3

The ortho position is one removed from the initial substituent's position. The meta position is two removed, and the para is three removed.

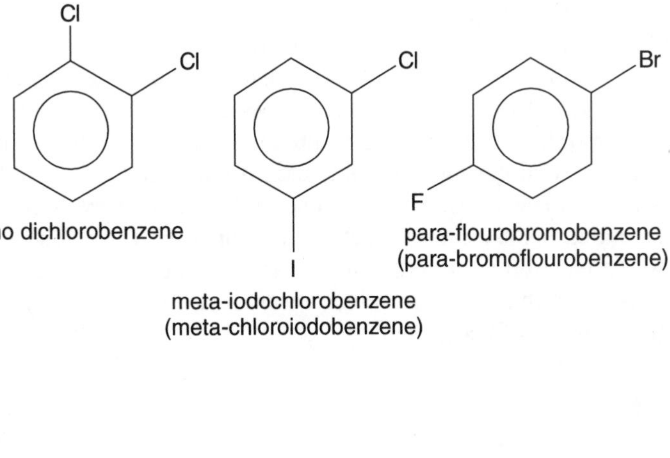

ortho dichlorobenzene

meta-iodochlorobenzene
(meta-chloroiodobenzene)

para-flourobromobenzene
(para-bromoflourobenzene)

Unlike the number system, you can assign an equally correct name with the names of the substituents reversed.

Benzene compounds that contain three or more substituents are always named by the number system. In this system, numbers are assigned to substituents so that the substituents have the lowest possible combination of numbers.

1-chloro-2-iodo-4-bromobenzene
(not 1-bromo-3-iodo-4-chlorobenzene)

Reactions of benzene

Although the resonance structures of benzene show it as a cyclo-hexatriene, because of its fully delocalized π system and the closed shell nature of this π system, benzene does not undergo addition reactions like ordinary unsaturated compounds. The destruction of the π electron system during addition reactions would make the products less stable than the starting benzene molecule. However, benzene does undergo substitution reactions in which the fully delocalized closed π electron system remains intact. For example, benzene may be reacted with a halogen in the presence of a Lewis acid (a compound capable of accepting an electron pair) to form a molecule of halobenzene.

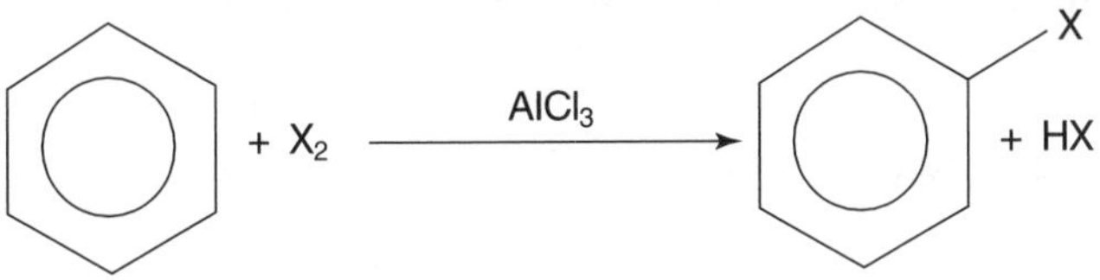

Other Aromatic Compounds

Many other compounds also exhibit aromatic characteristics. Some of the most common have two or more benzene rings fused together. Such compounds are called **polycyclic benzenoid aromatic compounds.** A typical example of this type of molecule is naphthalene, $C_{10}H_8$. Structurally, naphthalene looks like this:

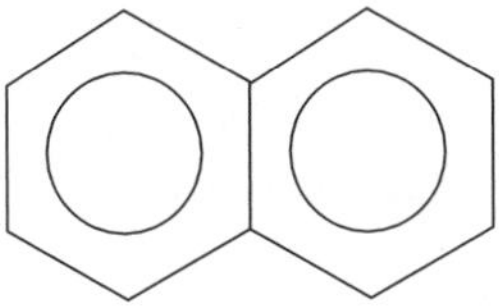

This structure is a hybrid of the following four resonance structures.

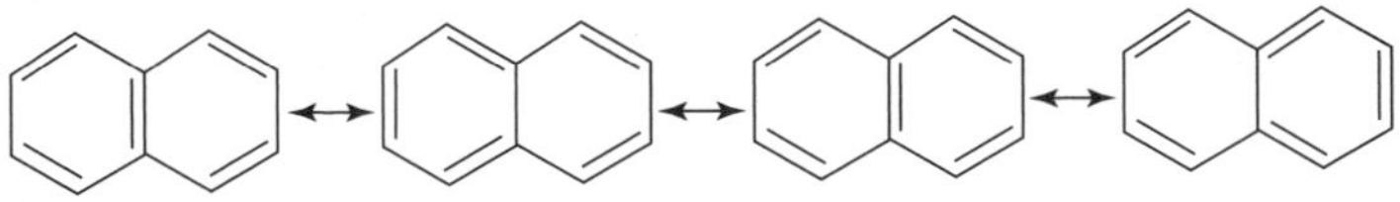

The resonance energy associated with naphthalene is 61 kcal/mole. Thus naphthalene is more stable than benzene. The benzenoid or benzene-like ring should have 36 kcal per mole of resonance energy, and the nonbenzenoid ring should have 25 kcal per mole of resonance energy. Because there is total delocalization of the electrons in the system, the benzenoid and nonbenzenoid rings cannot be identified at any given time.

Other benzenoid structures include anthracene, phenanthrene, and pyrene.

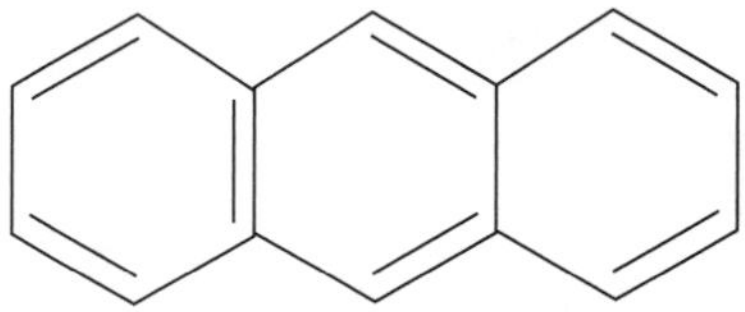

anthracene

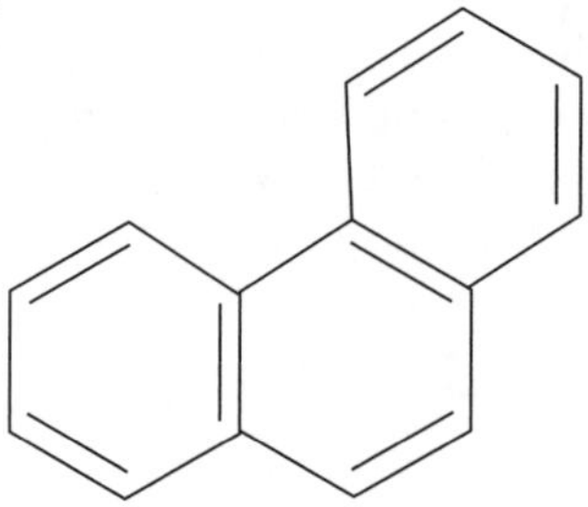

phenanthrene

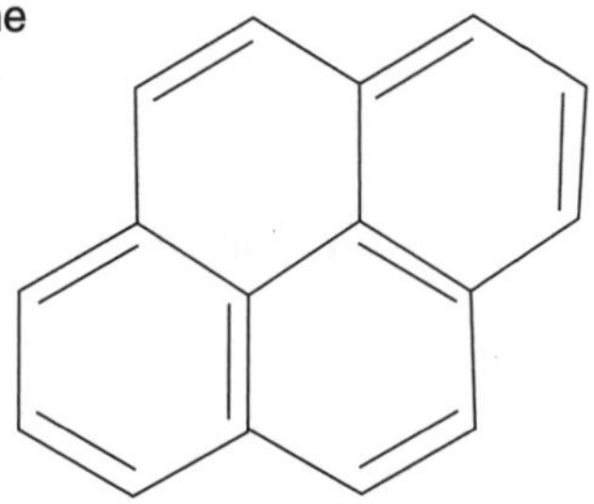

pyrene

Heterocyclic Aromatic Compounds

A **heterocyclic compound** is an organic compound in which one or more of the carbon atoms in the backbone of the molecule has been replaced by an atom other than carbon. Typical **hetero atoms** include nitrogen, oxygen, and sulfur.

Pyridine (C_5H_5N), pyrrole (C_4H_5N), furan (C_4H_4O), and thiophene (C_4H_4S) are examples of heteroaromatic compounds.

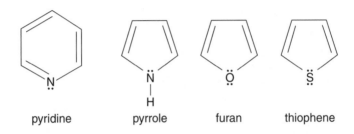

pyridine   pyrrole   furan   thiophene

Because these compounds are monocyclic aromatic compounds, they must obey Hückel's Rule. Hückel's Rule requires $4n + 2$ π electrons, so the simplest aromatic compound should contain 6 π electrons ($n = 1$). Pyrrole, furan, and thiophene appear, however, to have only 4 π electrons (2 π bonds). In systems such as these, the extra electrons needed to produce an aromatic condition come from the unshared electron pairs in sp^2 hybrid orbitals around the hetero atom.

Electrophilic Aromatic Substitution Reactions

Although aromatic compounds have multiple double bonds, these compounds do not undergo addition reactions. Their lack of reactivity toward addition reactions is due to the great stability of the ring systems that result from complete π electron delocalization (resonance). Aromatic compounds react by electrophilic aromatic substitution reactions, in which the aromaticity of the ring system is preserved. For example, benzene reacts with bromine to form bromobenzene.

Many functional groups can be added to aromatic compounds via electrophilic aromatic substitution reactions. A **functional group** is a substituent that brings with it certain chemical reactions that the aromatic compound itself doesn't display.

The bromination of benzene

All electrophilic aromatic substitution reactions share a common mechanism. This mechanism consists of a series of steps.

1. An **electrophile** — an electron-seeking reagent — is generated. For the bromination of benzene reaction, the electrophile is the Br+ ion generated by the reaction of the bromine molecule with ferric bromide, a Lewis acid.

$$Br\text{—}Br + FeBr_3 \longrightarrow Br\text{—}\overset{+}{Br}\text{—}FeBr_3^{-} \longrightarrow \overset{+}{Br} + FeBr_4^{-}$$

2. The electrophile attacks the π electron system of the benzene ring to form a nonaromatic carbocation.

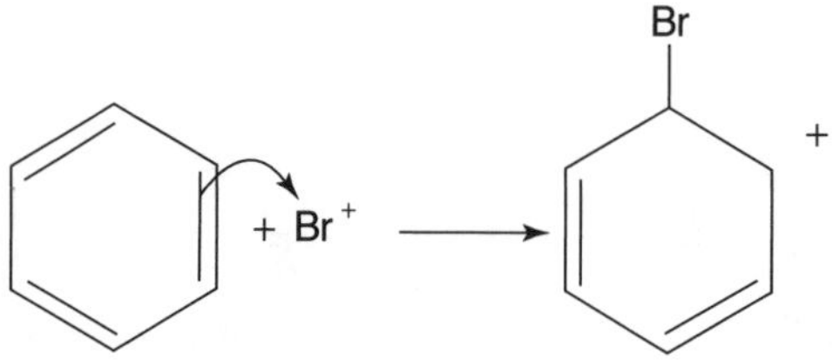

3. The positive charge on the carbocation that is formed is delocalized throughout the molecule.

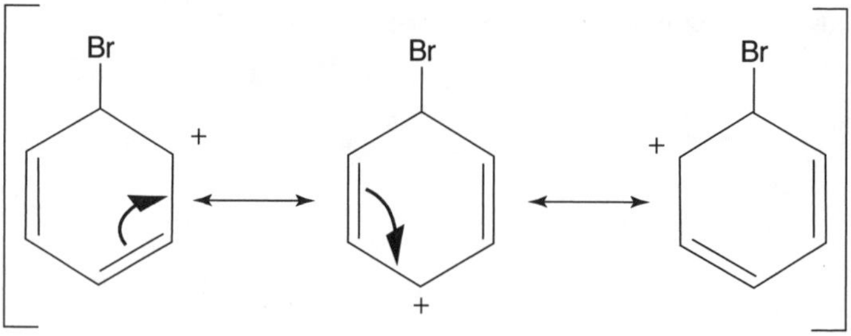

4. The aromaticity is restored by the loss of a proton from the atom to which the bromine atom (the electrophile) has bonded.

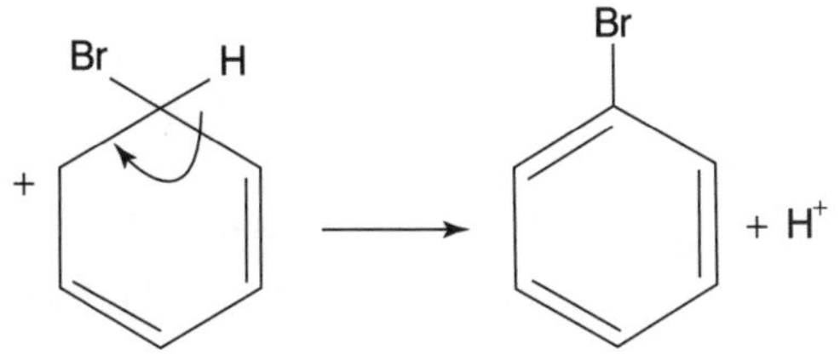

5. Finally, the proton reacts with the $FeBr_4^-$ to regenerate the $FeBr_3$ catalyst and form the product HBr.

$$H^+ + FeBr_4^- \longrightarrow FeBr_3 + HBr$$

You can summarize this particular electrophilic aromatic substitution mechanism like this:

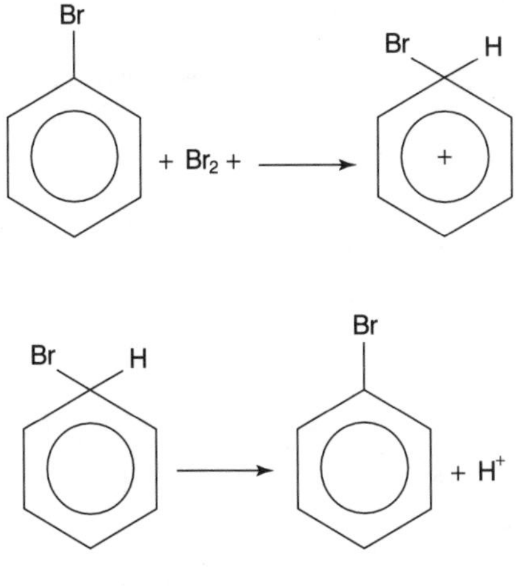

The nitration of benzene

In another example of an electrophilic aromatic substitution reaction, benzene reacts with a mixture of concentrated nitric and sulfuric acids to create nitrobenzene.

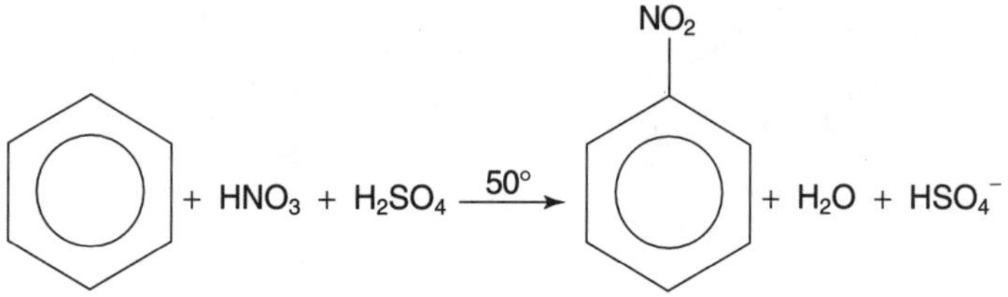

The mechanism for the nitrobenzene reaction occurs in six steps.

1. Sulfuric acid ionizes to produce a proton.

$$H_2SO_4 \rightleftharpoons H^+ + HSO_4^-$$

2. Nitric acid accepts the proton in an acid-base reaction.

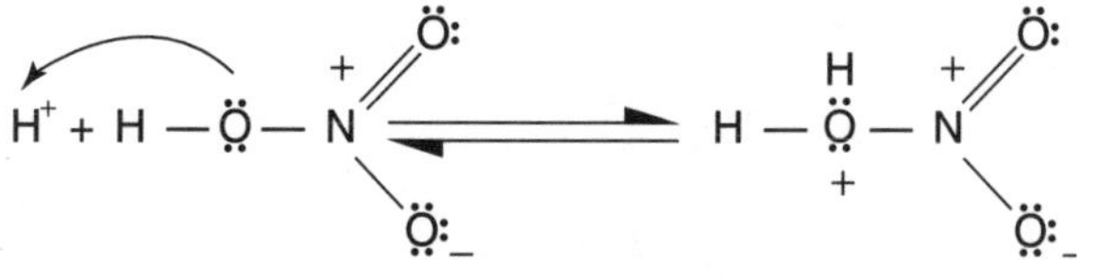

3. The protonated nitric acid dissociates to form a nitronium ion ($^+NO_2$).

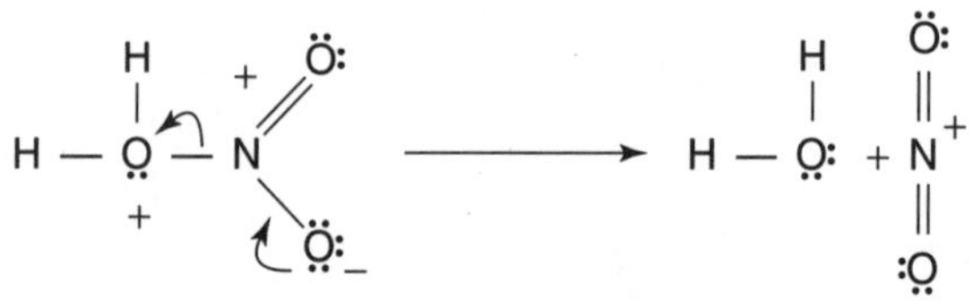

4. The nitronium ion acts as an electrophile and is attracted to the π electron system of the benzene ring.

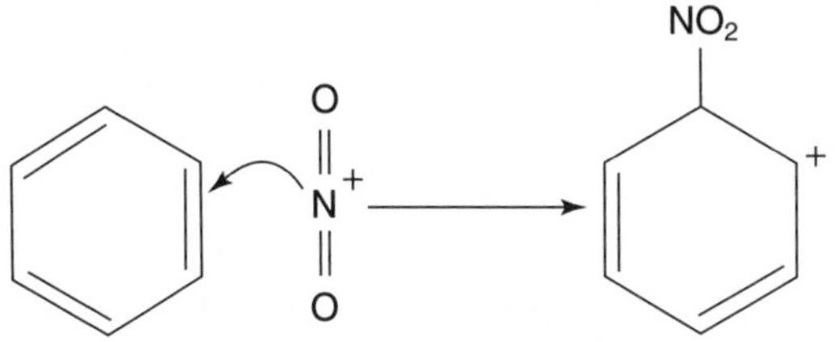

5. The nonaromatic carbocation that forms has its charge delocalized around the ring.

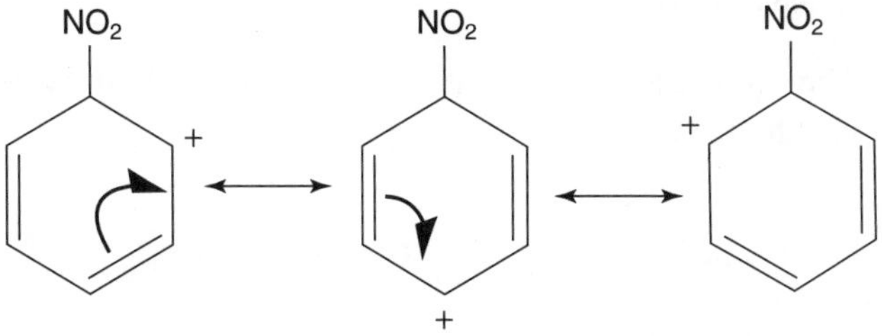

6. The aromaticity of the ring is reestablished by the loss of a proton from the carbon to which the nitro group is attached.

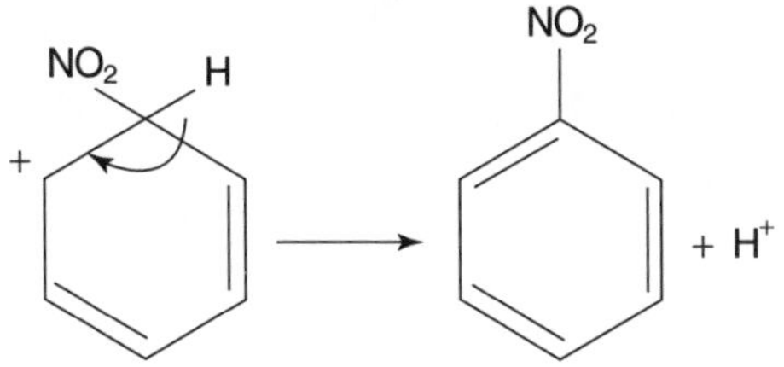

The sulfonation of benzene

The reaction of benzene with concentrated sulfuric acid at room temperature produces benzenesulfonic acid.

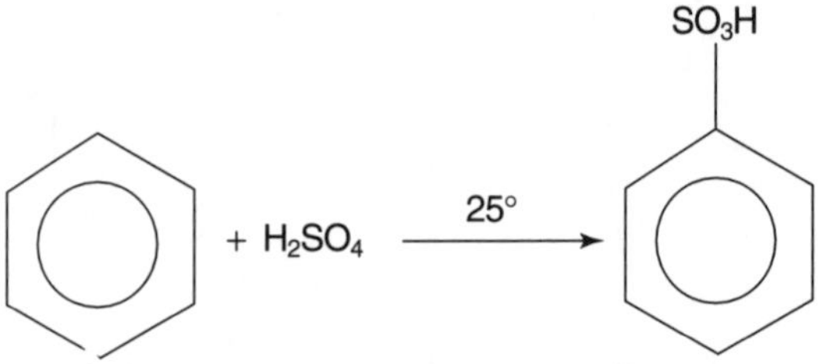

The mechanism for the reaction that produces benzenesulfonic acid occurs in the following steps:

1. The sulfuric acid reacts with itself to form sulfur trioxide, the electrophile.

$$2\ H_2SO_4 \rightleftharpoons SO_3 + H_3O^+ + HSO_4^-$$

This reaction takes place via a three-step process:

a. $H_2SO_4 \rightleftharpoons H^+ + HSO_4^-$

b.

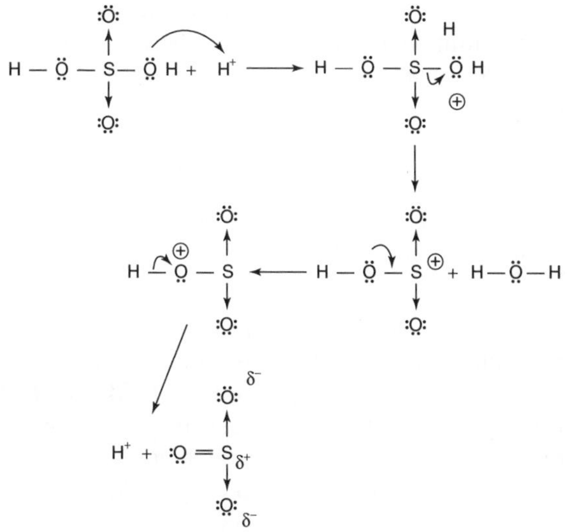

c. $H^+ + H_2O \longrightarrow H_3O^+$

2. The sulfur trioxide is attracted to the π electron system of the benzene molecule.

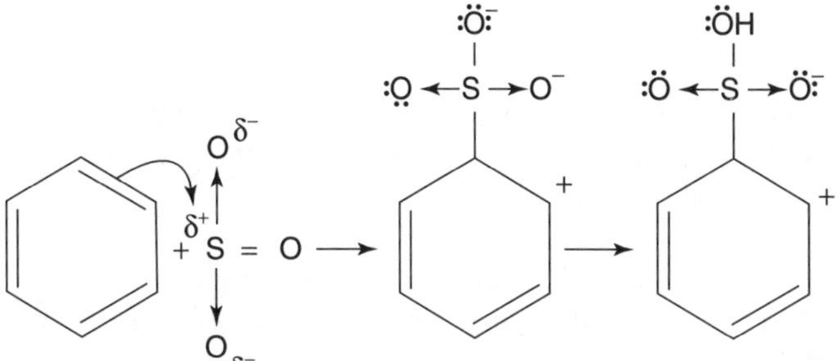

The remaining steps in the mechanism are identical with those in the bromination and nitration mechanisms: The charge around the ring is delocalized, and then the loss of a proton reestablishes the aromaticity of the ring.

The Birch Reduction of Benzene

The fully delocalized π electron system of the benzene ring remains intact during electrophilic aromatic substitution reactions. However, in the Birch reduction, this is not the case. In the **Birch reduction,** benzene, in the presence of sodium metal in liquid ammonia and methyl alcohol, produces a nonconjugated diene system. This reaction provides a convenient method for making a wide variety of useful cyclic dienes.

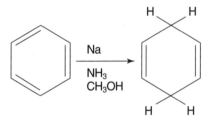

The production of the less stable nonconjugated diene instead of the more stable conjugated diene occurs because the reaction is kinetically controlled rather than thermodynamically controlled.

In general, reactions that aren't easily reversible are kinetically controlled because equilibrium is rarely established. In kinetically controlled reactions, the product with the lowest-energy transition state predominates. Reactions that are easily reversible are thermodynamically controlled, unless something occurs that prevents equilibrium. In thermodynamically controlled reactions, the lowest-energy product predominates.

Friedel-Crafts Alkylation Reaction

An alkyl group can be added to a benzene molecule by an electrophile aromatic substitution reaction called the **Friedel-Crafts alkylation reaction.** One example is the addition of a methyl group to a benzene ring.

The mechanism for this reaction begins with the generation of a methyl carbocation from methylbromide. The carbocation then reacts with the π electron system of the benzene to form a nonaromatic carbocation that loses a proton to reestablish the aromaticity of the system.

1.   An electrophile is formed by the reaction of methylchloride with aluminum chloride.

2. The electrophile attacks the π electron system of the benzene ring to form a nonaromatic carbocation.

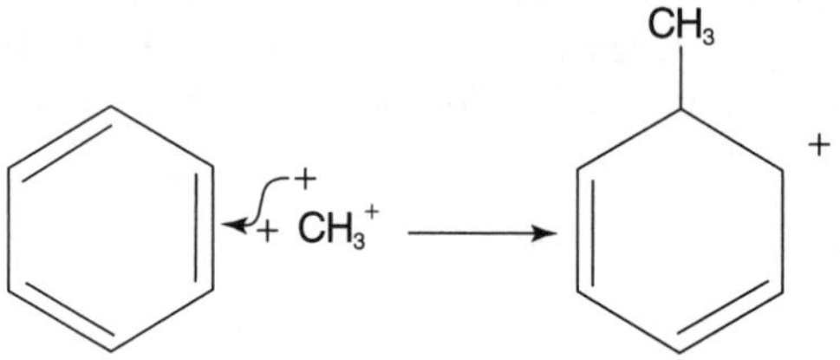

3. The positive charge on the carbocation that is formed is delocalized throughout the molecule.

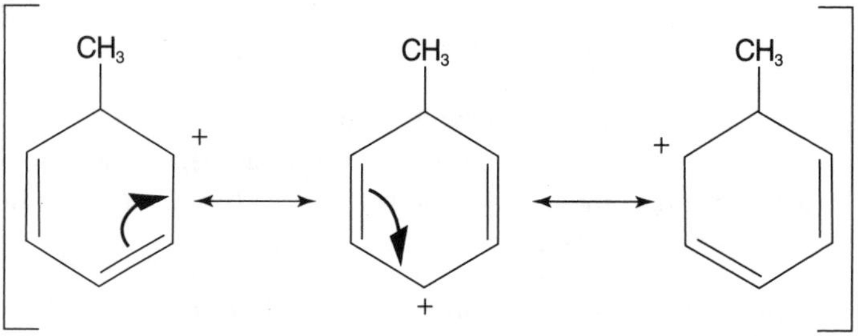

4. The aromaticity is restored by the loss of a proton from the atom to which the methyl group has bonded.

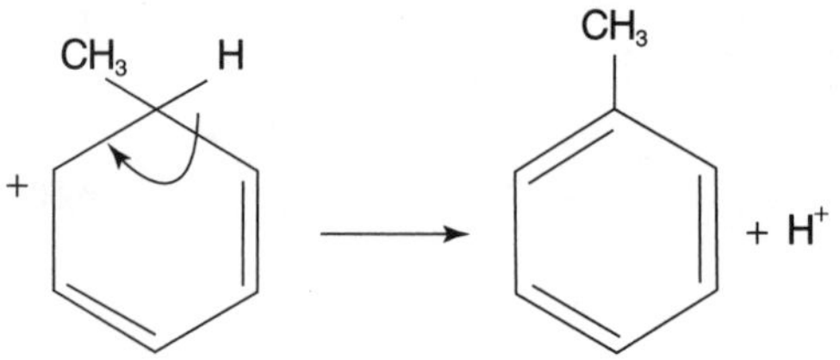

5. Finally, the proton reacts with the $AlCl_4^-$ to regenerate the $AlCl_3$ catalyst and form the product HCl.

$$H^+ + AlCl_4^- \longrightarrow HCl + AlCl_3$$

Carbocations can rearrange during the Friedel-Crafts alkylation reaction, leading to the formation of unpredicted products. One example is the formation of isopropyl benzene by the reaction of propyl chloride with benzene.

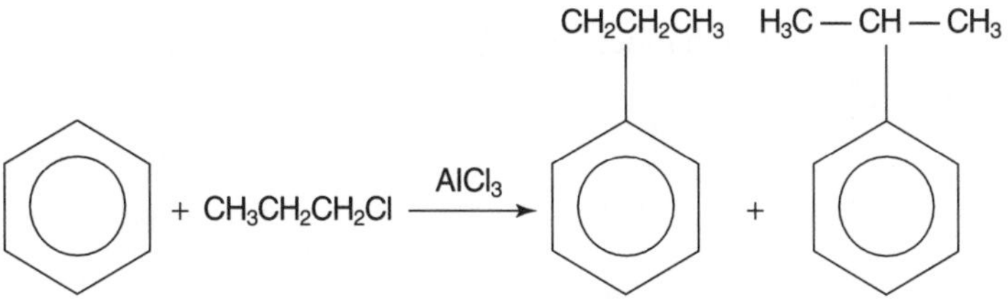

minor product   major product

The isopropyl benzene results from a rearrangement of the initially formed propyl carbocation to the more stable isopropyl carbocation.

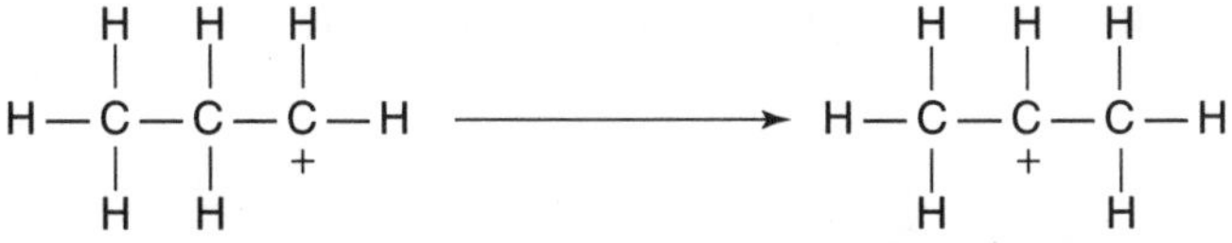

1° carbocation   2° carbocation

This rearrangement is called a 1,2-hydride ion shift. A hydride ion is H^-.

Friedel-Crafts Acylation Reaction

The **Friedel-Crafts acylation reaction,** another example of an electrophilic aromatic substitution reaction, is similar to the Friedel-Crafts alkylation reaction except that the substance that reacts with benzene is an acyl halide,

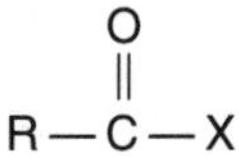

$$
\begin{array}{c}
\text{O} \\
\parallel \\
\text{R}-\text{C}-\text{X}
\end{array}
$$

instead of an alkyl halide, $R-X$. An acetyl chloride reaction appears as:

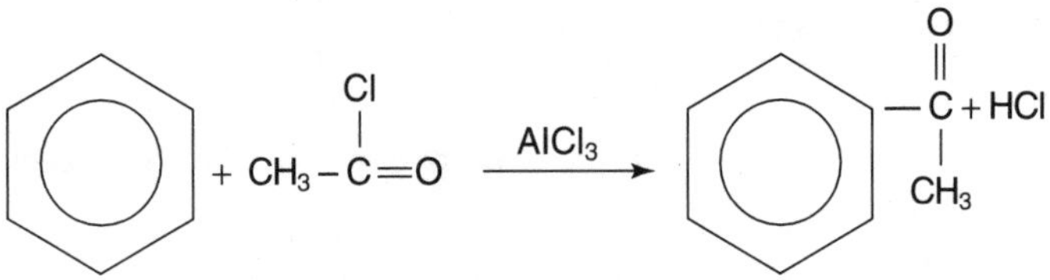

The mechanism for the generation of the acylium ion, $R-\overset{\overset{\textstyle O}{\parallel}}{C}{}^{+}$ is

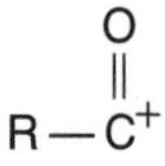

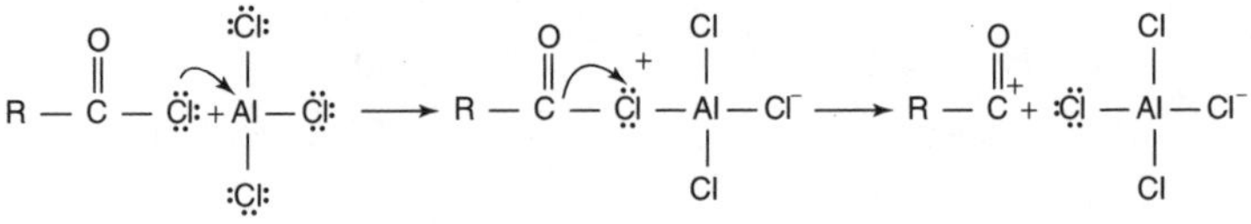

The remainder of the mechanism is identical to that of the alkylation of benzene. Because the acylium ion is resonance stabilized, no rearrangements occur.

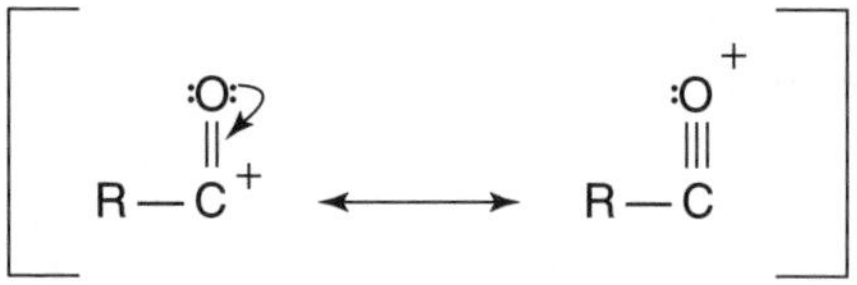

1. The reaction of acetyl chloride with aluminum chloride forms an electrophile.

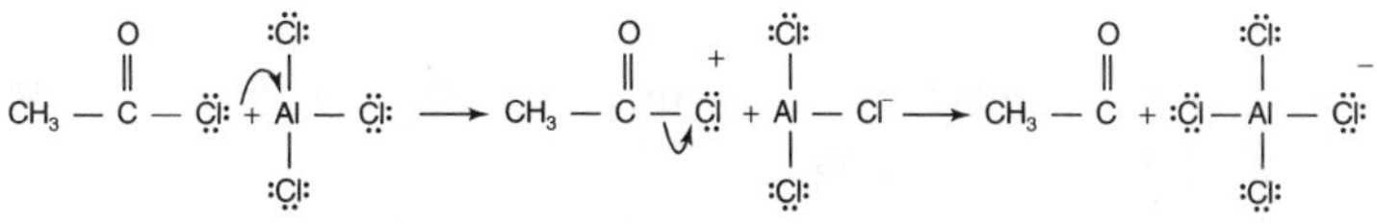

2. The electrophile attracts the π electron system of the benzene ring to form a nonaromatic carbocation.

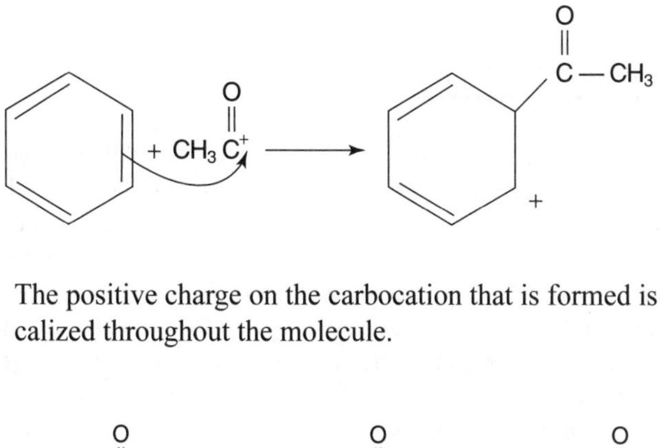

3. The positive charge on the carbocation that is formed is delocalized throughout the molecule.

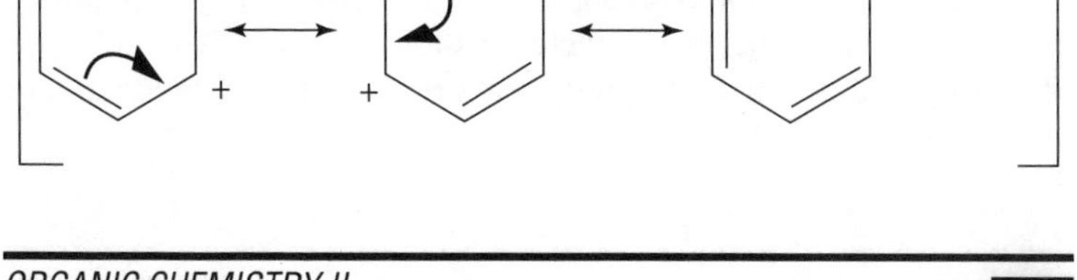

4. The loss of a proton from the atom to which the acetyl group has bonded restores the aromaticity.

5. The proton reacts with the $AlCl_4^-$ to regenerate the $AlCl_3$ catalyst and form the product HCl.

$$H^+ + AlCl_4^- \longrightarrow HCl + AlCl_3$$

Directing Group Influence

Substituents already attached to benzene exert an influence on additional atoms or groups attempting to bond to the benzene ring via electrophilic aromatic substitution reactions. An atom or group already attached to a benzene ring may direct an incoming electrophile to either the ortho-para positions or the meta position. Atoms or groups that make the benzene molecule more reactive by increasing the ring's electron density are called **activating groups.** Activating groups serve as ortho-para directors when they are attached to a benzene ring, meaning that they direct an incoming electrophile to the ortho or para positions. An atom or group that makes the benzene molecule less reactive by removing electron density from the ring acts as a **deactivating group.** Deactivating groups direct incoming electrophiles to the meta position. You can further classify activating and deactivating groups or atoms as strong, moderate, or weak in their directing influence. Table 2-1 lists some typical activating and deactivating groups by the order of their strength.

TABLE 2-1: The Influence of Various Directing Groups and Atoms

Ortho-Para Directors	Meta Directors
Strong	**Strong**
—NH$_2$	—NO$_2$
—NHR	—NH$_3^+$
—NR$_2$	—NR$_3^+$
—OH	—CF$_3$
—O$^-$	—CCl$_3$
Moderate	**Moderate**
—NHCOCH$_3$	—C≡N
—NHCOR	—SO$_3$H
—OCH$_3$	—COOH
—OR	—COOR
—CHO	
—COR	
Weak	**Weak**
—CH$_3$	
—CH$_2$CH$_3$	
—R	
—C$_6$H$_5$	

Halogen atom influence

Halogen atoms show both activating and deactivating characteristics. Because they have three pairs of unshared electrons, halogen atoms can supply electrons toward the ring. However, because of their high electronegativities, halogen atoms also tend to remove electrons from the benzene ring. These conflicting properties make halogens a weak ortho-para director and also a ring deactivator. This means that the presence of a halogen atom on a benzene causes an incoming electrophile to attach at an ortho or para position. However, these positions are not very electron rich, so the reaction proceeds poorly under ordinary electrophilic aromatic substitution conditions, leading to poor yields of the disubstituted product.

Predicting second group position

Using the previous information in this section, predicting the position that a second substituent group will take on a benzene ring is easy. For example, the nitration of methyl benzene (toluene) will produce ortho and para nitrotoluene as the main product because the methyl group is an ortho-para director.

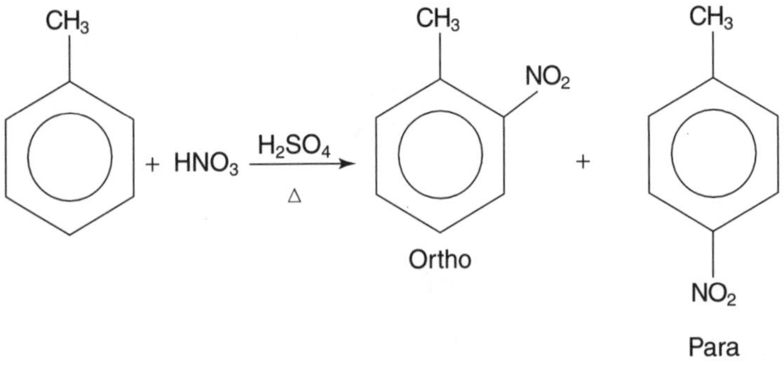

However, methylating nitrobenzene leads to the formation of meta nitrotoluene because the nitro group is meta directed.

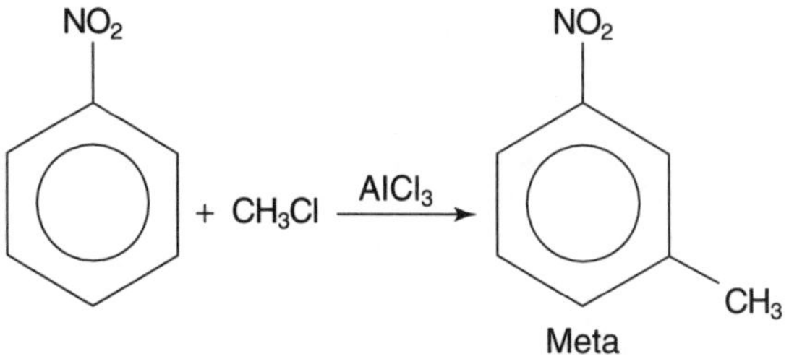

$$NO_2 \quad + \quad CH_3Cl \quad \xrightarrow{\text{AlCl}_3} \quad NO_2 \text{...} CH_3$$

Meta

Theory of Substitution Effects (Directing Group Influence)

Ring activators are groups that increase the electron density on the benzene ring and thereby make the ring more susceptible to electrophilic aromatic substitution reactions. **Ring deactivators** decrease the electron density on the benzene ring, thus making the ring less reactive toward electrophilic aromatic substitution reactions. Resonance theory can be used to illustrate these processes.

Ring Activation

Most ring activators have atoms with unshared electron pairs directly attached to a carbon atom of the benzene ring. For example, the — OH group has two pairs of unshared electrons on the oxygen atom, which will form a bond to a carbon atom of the benzene ring. Thus, the — OH group will be an activating group. The following illustration shows why this group will act as an ortho-para director.

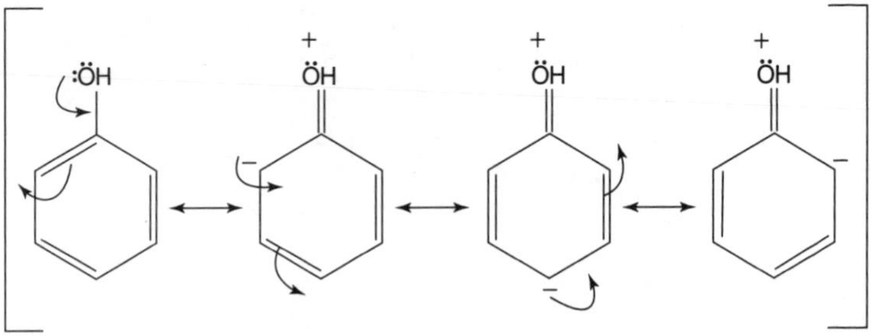

Notice that three of the four resonance structures show a negative charge residing on the positions ortho and para to the — OH group. These electron-rich positions should attract an electrophile more strongly than the less electron-rich meta positions do. Therefore, any group that possesses unshared electron pairs on the atom directly attached to a carbon atom of the benzene ring will be an ortho-para (activating) group. Groups that do not have unshared electron pairs on the atom directly attached to the benzene ring may also supply electrons to the benzene ring. This situation occurs if the atom in a group has weakly bonded π electrons attached to it or if the group has an inductive effect associated with it. The following diagram shows an example of the π electron movement giving ring activation.

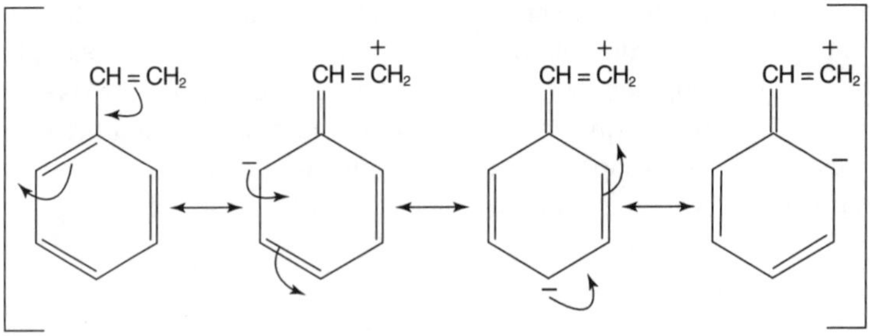

As with the — OH group example, the ortho and para positions are electron-rich compared to the meta positions. Thus ortho-para substitution occurs.

Ring Deactivation

Groups that withdraw electrons from the ring will deactivate the ring and act as meta directors. Groups capable of doing this usually contain an atom that is directly attached to a carbon atom of the benzene ring and that bears a positive or partially positive charge. A typical example is the nitro group — NO_2. The structure of the nitro group is:

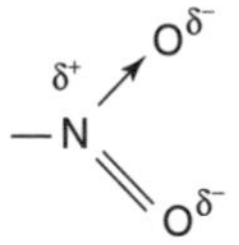

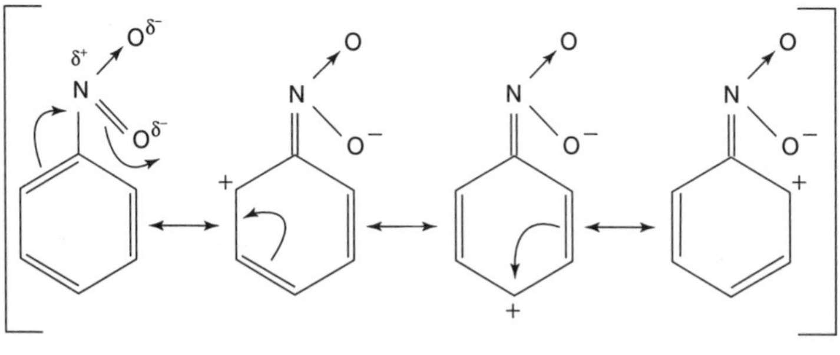

Notice that in three of the four resonance structures, a positive charge exists on the ortho and para positions. Thus, the hybrid structure is electron-poor in these areas, meaning that an electrophile generally attaches to the more electron-rich meta position.

Introduction

An **alkyl halide** is another name for a halogen-substituted alkane. The carbon atom, which is bonded to the halogen atom, has sp^3 hybridized bonding orbitals and exhibits a tetrahedral shape. Due to electronegativity differences between the carbon and halogen atoms, the σ covalent bond between these atoms is polarized, with the carbon atom becoming slightly positive and the halogen atom partially negative. Halogen atoms increase in size and decrease in electronegativity going down the family in the periodic table. Therefore, the bond length between carbon and halogen becomes longer and less polar as the halogen atom changes from fluorine to iodine.

Physical properties

Alkyl halides have little solubility in water but good solubility with nonpolar solvents, such as hexane. Many of the low molecular weight alkyl halides are used as solvents in reactions that involve nonpolar reactants, such as bromine. The boiling points of different alkyl halides containing the same halogen increase with increasing chain length. For a given chain length, the boiling point increases as the halogen is changed from fluorine to iodine. For isomers of the same compound, the compound with the more highly-branched alkyl group normally has the lowest boiling point. Table 3-1 summarizes data for some representative alkyl halides.

Table 3-1: Boiling Points (°C) of Alkyl Halides

	Flouride	*Chloride*	*Bromide*	*Iodine*
Group	**bp**	**bp**	**bp**	**bp**
Methyl	-78.4	-28.8	-3.6	42.5
Ethyl	-37.7	13.1	38.4	72
Propyl	-2.5	46.6	70.8	102
Isopropyl	-9.4	34	59.4	89.4
Butyl	32	78.4	101	130
Sec-butyl		68	91.2	120
Tert-butyl		51	73.3	100

Nomenclature

Alkyl halides are named using the IUPAC rules for alkanes. Naming the alkyl group attached to the halogen and adding the inorganic halide name for the halogen atom creates common names.

CH_3CH_2BR

bromoethane
(ethyl bromide)

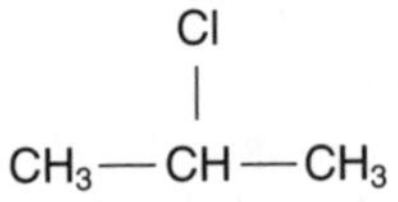

2-chloropropane
(isopropyl chloride)

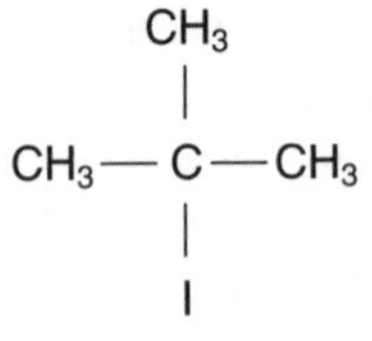

2-iodo-2-methylpropane
(tertiary-butyliodide)

Nucleus and Nucleophiles

A **nucleus** is any atom that has a partial or fully positive charge associated with it. A **nucleophile** is an atom or group that is attracted to a source of partial or full positive charge. Alkyl halides act as a nucleus because of the great electronegativity differences between the carbon atom and the halogen atom directly bonded to it. This great electronegativity difference causes the electron density in the overlap region between the carbon and halogen atoms to be pulled toward the halogen atom. This shifting of electron density in the molecule makes the carbon atom partially positive (the nucleus) and the halide ion partially negative (the incipient leaving group).

Figure 3-1 illustrates the effect of electronegativity differences on bond polarity.

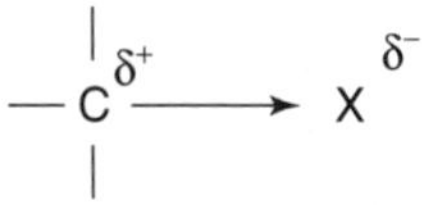

$$-\overset{\displaystyle |}{\underset{\displaystyle |}{C}}\,{}^{\delta^+}\longrightarrow\ X\ {}^{\delta^-}$$

Figure 3-1

Electrons in the overlap region between the carbon and the halogen atoms are attracted to the more electronegative halogen atom. The carbon atom, which now has less of a share of the bonding electrons, becomes partially positive, and the halogen atom, which has a greater share of these electrons, becomes partially negative.

Remember that a nucleophile is a substance that has a pair of electrons that it can donate to another atom. The weaker the forces of attraction holding the electron pair to the original molecule, the more readily this molecule will share the electrons and the stronger the resulting nucleophile will be. The weakest held electrons on an atom are the nonbonding electron pairs. Electrons in π bonds, although held more strongly than nonbonding electrons, are also loosely held and easily shared, making unsaturated compounds relatively good nucleophiles.

Because they possess a negative charge, anions are always better nucleophiles than their conjugate acids.

Nucleophilic Substitution Reactions

Alkyl halides undergo many reactions in which a nucleophile displaces the halogen atom bonded to the central carbon of the molecule. The displaced halogen atom becomes a halide ion.

$$Nu^- \quad + \quad R-X \quad \longrightarrow \quad R-Nu \quad + \quad :\ddot{X}:^-$$

nucleophile   alkyl halide   product   halide ion

Some typical nucleophiles are the hydroxy group (^-OH), the alkoxy group (RO^-), and the cyanide ion ($^-C{\equiv}N$). Reaction of these nucleophiles with an alkyl halide (R—X) gives the following reactions and products:

$:\overset{-}{\underset{}{O}}H$ + R−X ⟶ R−OH + $:\overset{..}{\underset{..}{X}}:^{-}$

hydroxide ion   alkyl halide   alcohol   halogen ion

$R\overset{..}{\underset{..}{O}}:^{-}$ + R−X ⟶ R−O−R + $:\overset{..}{\underset{..}{X}}:^{-}$

aldoxide ion   alkyl halide   ether   halogen ion

$^{-}:C\equiv N$ + R−X ⟶ R−C≡N + $:\overset{..}{\underset{..}{X}}:^{-}$

cyanide ion   alkyl halide   nitrile   halogen ion

$R−\overset{..}{N}H_2$ + R−X ⟶ RNHR + $:\overset{..}{\underset{..}{X}}:^{-}$

primary amine   alkyl halide   2° amine   halogen ion

The halogen ion that is displaced from the carbon atom is called the **leaving group,** and the overall reaction is called a **nucleophilic substitution reaction.**

Leaving Group

For a molecule to act as a nucleus or substrate in a nucleophilic substitution reaction, it must have both a polar bond and a good leaving group. For an atom or a group to be a good leaving group, it must be able to exist independently as a relatively stable, weakly basic ion or molecule. Groups that act as leaving groups are always capable of accommodating the negative charge through a high electronegativity or by delocalization. Because halogen atoms have high electronegativities and form relatively stable ions, they act as good leaving groups.

Mechanisms of Nucleophilic Substitution Reactions

Experimental data from nucleophilic substitution reactions on substrates that have **optical activity** (the ability to rotate plane-polarized light) shows that two general mechanisms exist for these types of reactions. The first type is called an S_N2 mechanism. This mechanism follows **second-order kinetics** (the reaction rate depends on the concentrations of two reactants), and its intermediate contains both the substrate and the nucleophile and is therefore bimolecular. The terminology S_N2 stands for "substitution nucleophilic bimolecular."

The second type of mechanism is an S_N1 mechanism. This mechanism follows **first-order kinetics** (the reaction rate depends on the concentration of one reactant), and its intermediate contains only the substrate molecule and is therefore unimolecular. The terminology S_N1 stands for "substitution nucleophilic unimolecular."

S_N2 mechanism

As noted earlier in this chapter, the alkyl halide substrate contains a polarized carbon halogen bond. The S_N2 mechanism begins when an electron pair of the nucleophile attacks the back lobe of the leaving group. Carbon in the resulting complex is trigonal bipyramidal in shape. With the loss of the leaving group, the carbon atom again assumes a pyramidal shape; however, its configuration is inverted. See Figure 3-2 below.

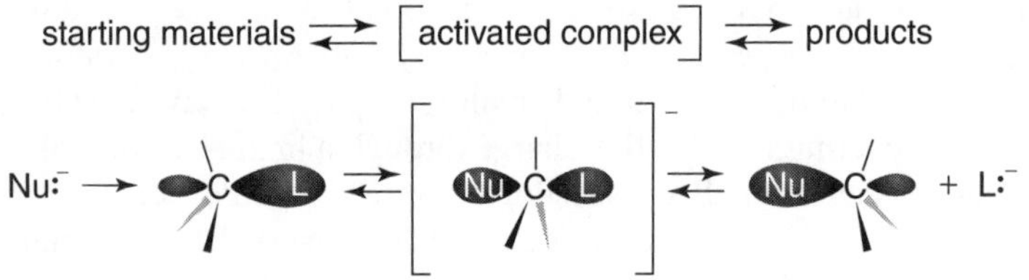

Figure 3-2

The S$_N$2 mechanism can also be illustrated as shown in Figure 3-3.

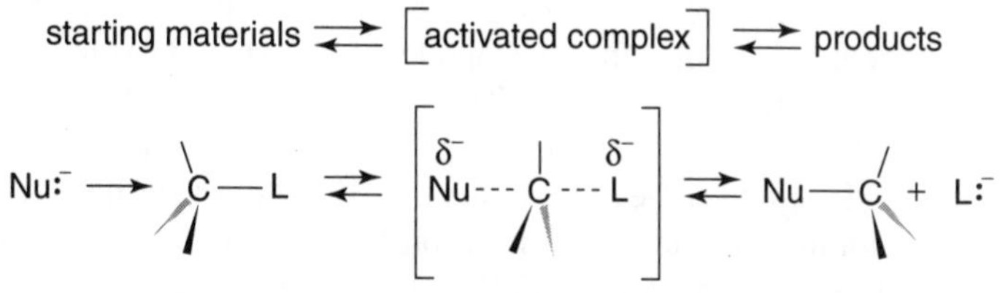

starting materials ⇌ [activated complex] ⇌ products

$$Nu: \longrightarrow \quad \underset{|}{\overset{\diagdown}{C}}-L \quad \rightleftharpoons \quad \left[\overset{\delta^-}{Nu} - - \underset{|}{\overset{|}{C}} - - \overset{\delta^-}{L} \right] \quad \rightleftharpoons \quad Nu-\overset{/}{C} + L:^-$$

Figure 3-3

Notice that in either picture, the intermediate shows both the nucleophile and the substrate. Also notice that the nucleophile must always attack from the side opposite the side that contains the leaving group. This occurs because the nucleophilic attack is always on the back lobe (antibonding orbital) of the carbon atom acting as the nucleus.

S$_N$2 mechanisms always proceed via rearward attack of the nucleophile on the substrate. This process results in the inversion of the relative configuration, going from starting material to product. This inversion is often called the **Walden inversion,** and this mechanism is sometimes illustrated as shown in Figure 3-4.

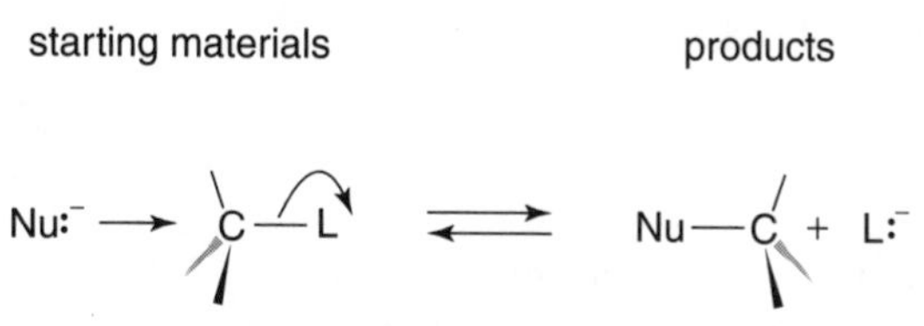

starting materials   products

$$Nu: \longrightarrow \quad \underset{|}{\overset{\diagdown}{C}}-L \quad \rightleftharpoons \quad Nu-\overset{/}{C} + L:^-$$

Figure 3-4

Steric hindrance

S_N2 reactions require a rearward attack on the carbon bonded to the leaving group. If a large number of groups are bonded to the same carbon that bears the leaving group, the nucleophile's attack should be hindered and the rate of the reaction slowed. This phenomenon is called **steric hindrance.** The larger and bulkier the group(s), the greater the steric hindrance and the slower the rate of reaction. Table 3-2 shows the effect of steric hindrance on the rate of reaction for a specific, unspecified nucleophile and leaving group. Different nucleophiles and leaving groups would result in different numbers but similar patterns of results.

Table 3-2: Effects of Steric Hindrance upon Rates of S_N2 Reactions

$$Nu^- + ALK - L \longrightarrow Nu - ALK + L^-$$

Alkyl Group (ALK)	Relative Rate of Substitution
-CH₃ (small group)	30
-CH₂CH₃ (larger group)	1
-CH(CH₃)₂ (bulky group)	0.03
-C(CH₃)₃ (very bulky group)	0

S_N2 reactions give good yields on 1° (primary) alkyl halides, moderate yields on 2° (secondary) alkyl halides, and poor to no yields on 3° (tertiary) alkyl halides.

Solvent effects

For **protic solvents** (solvents capable of forming hydrogen bonds in solution), an increase in the solvent's polarity results in a decrease in the rate of S_N2 reactions. This decrease occurs because protic solvents solvate the nucleophile, thus lowering its ground state energy. Because the energy of the activated complex is a fixed value, the energy of activation becomes greater and, therefore, the rate of reaction decreases.

Polar **aprotic solvents** (solvents that cannot form hydrogen bonds in solution) do not solvate the nucleophile but rather surround the accompanying cation, thereby raising the ground state energy of the nucleophile. Because the energy of the activated complex is a fixed value, the energy of activation becomes less and, therefore, the rate of reaction increases.

Figure 3-5 illustrates the effect of solvent polarity on the energy of activation and, thus, the rate of reaction.

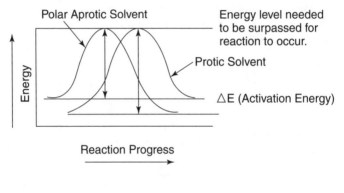

Figure 3-5

The smaller activation energy leads to the more rapid reaction.

S_N1 mechanism

The second major type of nucleophilic substitution mechanism is the S_N1 mechanism. This mechanism proceeds via two steps. The first step (the slow step) involves the breakdown of the alkyl halide into an alkyl carbocation and a leaving group anion. The second step (the fast step) involves the formation of a bond between the nucleophile and the alkyl carbocation.

$$\text{ALK} - \text{L} \xrightarrow[\text{step}]{\text{slow}} \left[\boxed{\text{Alkyl}} \right]^{+} \xrightarrow[\text{fast step}]{\text{Nu}} \text{ALK} - \text{Nu}$$

$$+\text{L}^{-}$$

Because the activated complex contains only one species—the alkyl carbocation—the substitution is considered unimolecular.

Carbocations contain sp² hybridized orbitals and thus have planar structures. S_N1 mechanisms proceed via a carbocation intermediate, so a nucleophile attack is equally possible from either side of the plane. Therefore, a pure, optically active alkyl halide undergoing an S_N1 substitution reaction will generate a racemic mixture as a product, as shown in Figure 3-6.

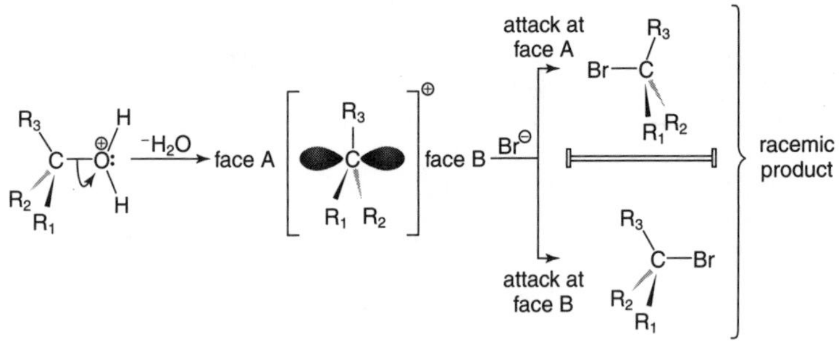

Figure 3-6

S_N1 versus S_N2 Reactions

Whether an alkyl halide will undergo an S_N1 or an S_N2 reaction depends upon a number of factors. Some of the more common factors include the natures of the carbon skeleton, the solvent, the leaving group, and the nature of the nucleophile.

Nature of the carbon skeleton

Only those molecules that form extremely stable cations undergo S_N1 mechanisms. Normally, only compounds that yield 3° (tertiary) carbonications (or resonance-stabilized carbocations) undergo S_N1 mechanisms rather than S_N2 mechanisms. Carbocations of tertiary alkyl halides not only exhibit stability due to the inductive effect, but the original molecules exhibit steric hindrance of the rear lobe of the bonding orbital, which inhibits S_N2 mechanisms from occurring. Primary alkyl halides, which have little inductive stability of their cations and exhibit no steric hindrance of the rear lobe of the bonding orbital, generally undergo S_N2 mechanisms. Figure 3-7 illustrates the tendencies of alkyl halides toward the two types of substitution mechanisms.

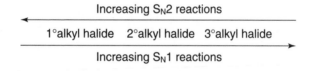

Figure 3-7

Nature of the solvent

Polar protic solvents such as water favor S_N1 reactions, which produce both a cation and an anion during reaction. These solvents are capable of stabilizing the charges on the ions formed during solvation. Because S_N2 reactions occur via a **concerted mechanism** (a mechanism which takes place in one step, with bonds breaking and forming at the same time) and no ions form, polar protic solvents would have little effect upon them. Solvents with low dielectric constants tend not to stabilize ions and thus favor S_N2 reactions. Conversely, solvents of high dielectric constants stabilize ions, favoring S_N1 reactions.

Nature of the leaving group

In general, good leaving groups are those capable of forming stable ions or molecules upon displacement from the original molecule. Conversely, poor leaving groups form ions of poor to moderate stability. Strong bases, such as OH¯, NH₂¯, and RO¯, make poor leaving groups. Water, which is less basic than a hydroxide ion, is a better leaving group. Poor bases usually make good leaving groups. A poor base is an ion or group in which the electrons are tightly bound to the molecule due to high electronegativity or resonance. Some good leaving groups are the sulfate ion and the p-toluenesulfonate (tosylate ion).

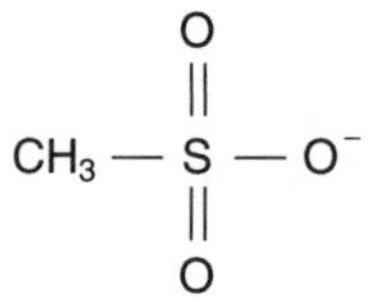

methyl sulfonate
(resonance stabilized)

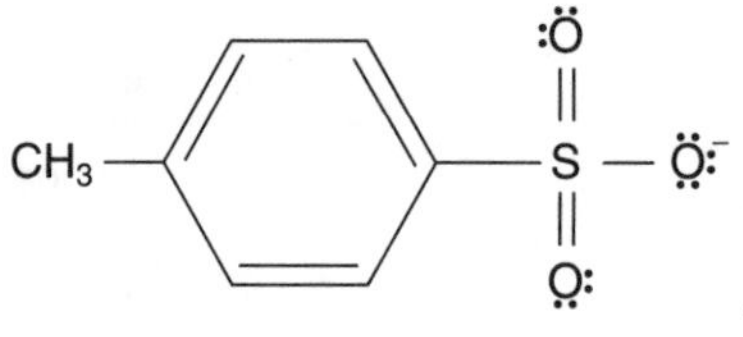

tosylate ion
(resonance stabilized)

The following list ranks atoms and molecules in order of their stability as leaving groups, from most to least stable.

$$CH_3SO_3^-, \text{ tosyl}^- > I^- > Br^- > H_2O^+ > Cl^- > F^-$$

Elimination Reactions

During an **elimination reaction,** a bond forms by the removal of two atoms or groups from the original molecule. In most instances, the bond that forms is a π bond. Elimination reactions compete with substitution reactions when alkyl halides react with a nucleophile.

$$CH_3CH_2Cl \ + \ R\ddot{\underset{..}{O}}:^- \longrightarrow CH_3CH_2\ddot{\underset{..}{O}}R \ + \ :\ddot{\underset{..}{C}}l^- \ \text{(substitution reaction)}$$

alkyl halide   nucleophile   ether   chloride ion

$$CH_3CH_2Cl \ + \ R\ddot{\underset{..}{O}}:^- \longrightarrow CH_2{=\!=}CH_2 \ + \ R\ddot{\underset{..}{O}}H \ + \ HCl \ \text{(elimination reaction)}$$

alkyl halide   base   alkene   alcohol   halogen acid

The elimination of hydrogen halide (a halogen acid) from an alkyl halide requires a strong base such as the alkoxide ion, RO. Weaker bases such as the OH ion give poor yields of elimination product.

If an alkyl halide contains more than two carbons in its chain, and the carbon atoms adjacent to the carbon atom bonded to the halogen each have hydrogen atoms bonded to them, two products will form. The major product is predicted by **Zaitsev's Rule,** which states that the more highly branched alkene will be the major product. For example, in the dehydrohalogenation reaction between 2-chlorobutane and sodium methoxide, the major product is 2-butene.

1-butene   2-butene

Mechanism of Elimination Reactions

As noted earlier, the halogen-carbon bond in an alkyl halide is polarized due to the electronegativity difference between the atoms. This polarization can lead to the formation of a partial or fully positive charge on the carbon atom.

The full or partial positive charge on the carbon atom is delocalized (dispersed) down the carbon chain. This, in turn, makes the hydrogen atoms attached to these carbons very slightly positive and thus very weakly acidic. Therefore, a very strong base can now remove a slightly positive hydrogen with the resulting release of electrons down the chain, forming a π bond between the carbon atoms. The actual mechanism can be one of two types, E1 or E2, depending upon the structure of the activated complex.

E1 mechanism

An atom that bears a pair of unshared electrons takes on one of two roles. The atom may share these electrons with a carbon atom that bears a leaving group, or it may share these electrons with a hydrogen atom. In the former case, the atom acts as a nucleophile, while in the latter case it acts as a base. Therefore, depending on reaction conditions, the atom may be involved in a substitution reaction or an elimination reaction.

The reaction of an OH ion with tertiary butyl bromide leads to little or no substitution product because steric hindrance blocks the rear lobe of the carbon atom to which the bromine atom is bonded. With the aid of a polar solvent, the bromine-carbon bond ionizes to form a tertiary carbocation and a bromide ion. The hydrogen atoms on the carbons adjacent to the carbocation carbon acquire a slight positive charge, allowing the OH⁻ ion to employ its basic characteristics. Thus, the OH⁻ ion abstracts a hydrogen atom, and the electrons migrate down the chain, forming a double bond.

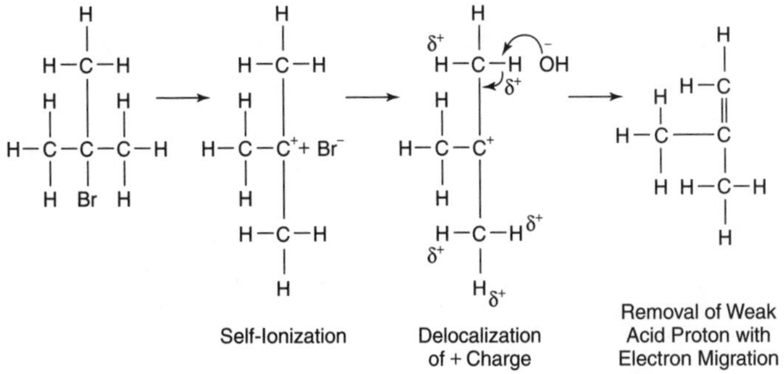

| Self-Ionization | Delocalization of + Charge | Removal of Weak Acid Proton with Electron Migration |

The activated complex for this reaction contains only the alkyl halide and is, therefore, unimolecular. The reaction follows an E1 mechanism.

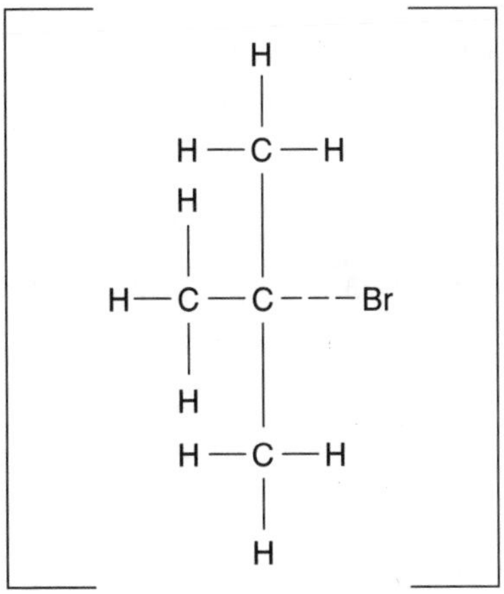

E2 mechanism

Elimination reactions can also occur when a carbon halogen bond does not completely ionize, but merely becomes polarized. As with the E1 reactions, E2 mechanisms occur when the attacking group displays its basic characteristics rather than its nucleophilic property. The activated complex for this mechanism contains both the alkyl halide and the alkoxide ion.

Following is the complete mechanism for the E2 elimination reaction:

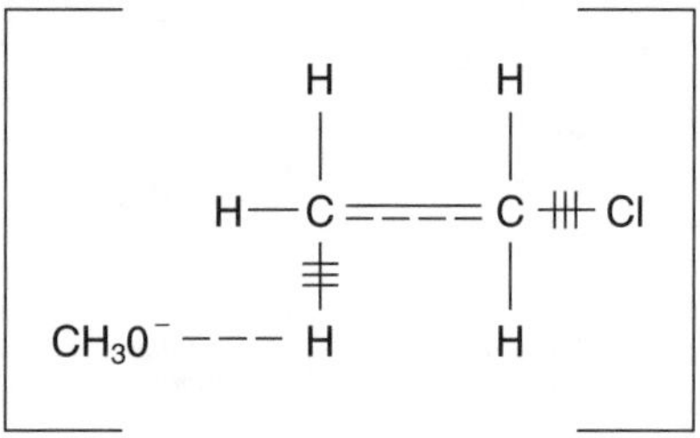

Grignard Reaction

In a **Grignard reaction,** an alkyl halide reacts with magnesium metal in an anhydrous ether solvent to create an organometallic reagent.

$$CH_3Br \quad + \quad Mg \quad \xrightarrow{\text{ether}} \quad CH_3MgBr$$

methyl bromide   magnesium   methyl magnesium bromide

The Grignard reagent is highly reactive and is used to prepare many functional groups. An example is the preparation of a carboxylic acid by reaction with carbon dioxide and mineral acid.

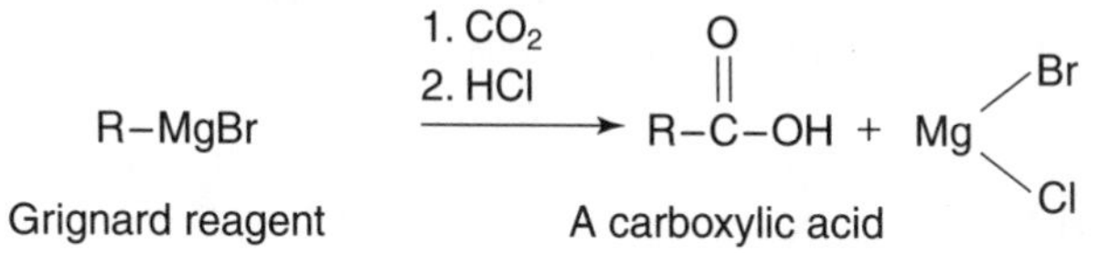

$$\text{R－MgBr} \xrightarrow[\text{2. HCl}]{\text{1. CO}_2} \quad \overset{\overset{\text{O}}{\|}}{\text{R－C－OH}} + \text{Mg}\overset{\text{Br}}{\underset{\text{Cl}}{\diagdown}}$$

Grignard reagent   A carboxylic acid

Preparation of Alkyl Halides

Following are two methods commonly used to prepare alkyl halides.

Hydrogen halide addition to an alkene

Halogen halides add across carbon-carbon double bonds. These additions follow Markovnikov's rule, which states that the positive part of a reagent (a hydrogen atom, for example) adds to the carbon of the double bond that already has more hydrogen atoms attached to it. The negative part adds to the other carbon of the double bond. Such an arrangement leads to the formation of the more stable carbocation over other less-stable intermediates.

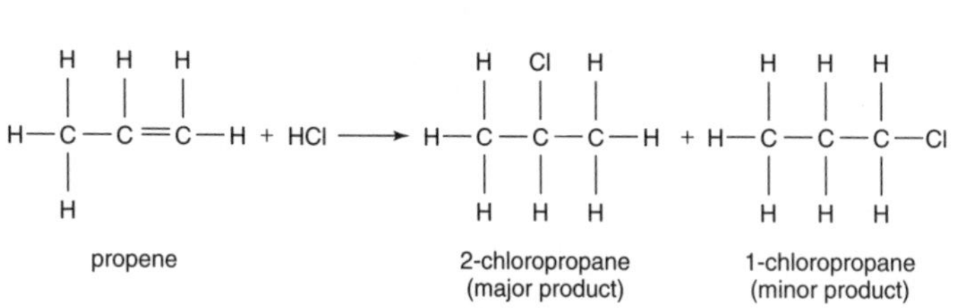

propene   2-chloropropane (major product)   1-chloropropane (minor product)

Reaction of alcohols with sulfur and phosphorous halides

Alcohols can be converted to alkyl halides by reaction with thionyl chloride, $SOCl_2$; phosphorous trichloride, PCl_3; phosphorous pentachloride, PCl_5; or phosphorous tribromide, PBr_3. For example, ethyl chloride or ethyl bromide can be prepared from ethyl alcohol via reactions with sulfur and phosphorous halides.

$$CH_3CH_2OH \ + \ SOCl_2 \ \xrightarrow{\Delta} \ CH_3CH_2Cl \ + \ SO_2 \ + \ HCl$$

$$CH_3CH_2OH \ + \ PCl_3 \ \xrightarrow{\Delta} \ CH_3CH_2Cl \ + P(OH)_3 + \ HCl$$

$$CH_3CH_2OH \ + \ PCl_5 \ \xrightarrow{\Delta} \ CH_3CH_2Cl \ + POCl_3 + \ HCl$$

$$CH_3CH_2OH \ + \ PBr_3 \ \xrightarrow{\Delta} \ CH_3CH_2Br \ + P(OH)_3 + \ HBr$$

Phenols

Phenols are molecules that have a hydroxyl group attached to the carbon atom of an aromatic ring.

Nomenclature

By definition, phenol is hydroxybenzene. Phenol is a common name for the compound. Its IUPAC name would be benzenol, derived in the same manner as the IUPAC names for aliphatic alcohols.

When a phenol molecule is substituted with additional groups, either the ortho, meta, para system or the numbering system can be employed. In either case, if the parent molecule is referred to as a *phenol*, the nomenclature being used is the common system.

In IUPAC nomenclature, the parent molecule is called benzenol, and substituents are always numbered with the OH group being given the understood first position. For the compounds below, the first name listed is the common name and the second is the IUPAC name.

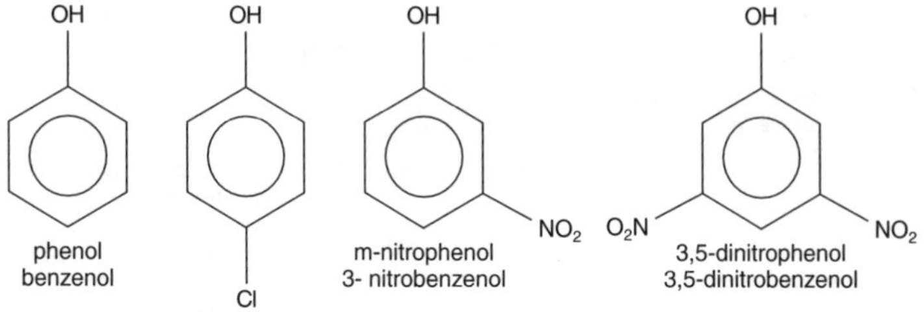

phenol
benzenol

p-chlorophenol
4-chlorobenzenol

m-nitrophenol
3- nitrobenzenol

3,5-dinitrophenol
3,5-dinitrobenzenol

Certain phenols are referred to by common names. For example, methyl phenols are called *cresols*. In the illustrations below, the first name under each compound is its common name, and the bottom name is its IUPAC name.

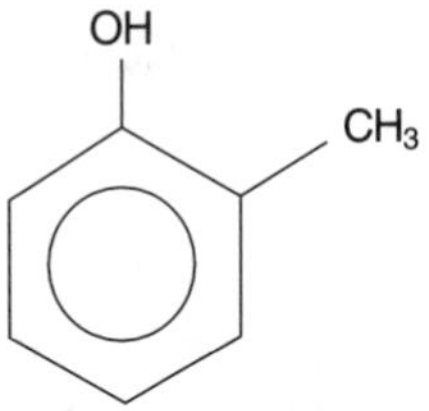

2-methylphenol (o-cresol)
2-methylbenzenol

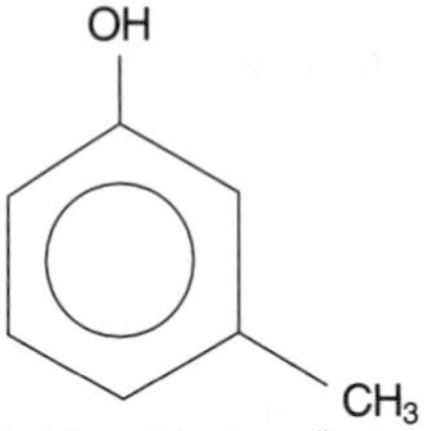

3-methylphenol (m-cresol)
3-methylbenzenol

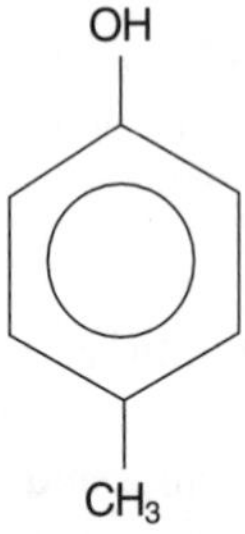

4-methylphenol (p-cresol)
4-methylbenzenol

Similarly, hydroxyphenols have common names, which are listed first under each of the following illustrations, while the IUPAC names are listed last.

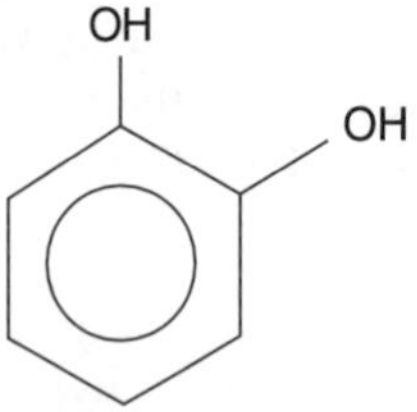

2-hydroxyphenol (catechol)
1,2-benzenediol

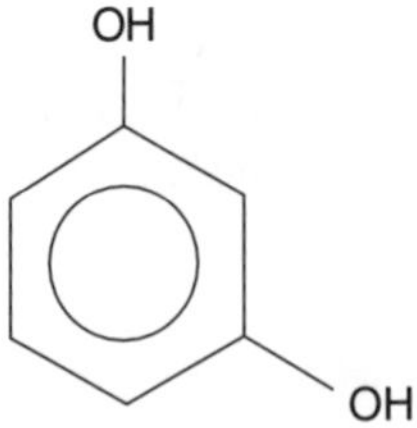

3-hydroxyphenol (resorcinol)
1,3-benzenediol

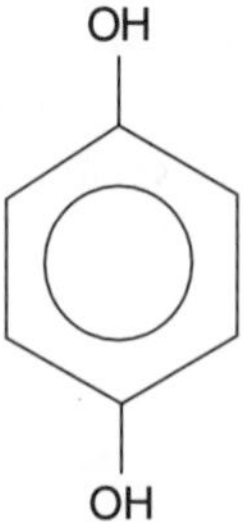

4-hydroxyphenol (hydroquinone)
1,4-benzenediol

Physical properties

Low molecular weight phenols are normally liquids or low melting solids. Due to hydrogen bonding, most low molecular weight phenols are water-soluble. Phenols tend to have higher boiling points than alcohols of similar molecular weight because they have stronger intermolecular hydrogen bonding.

Acidity

Phenols show appreciable acidity ($pK_a = 8^{-10}$). For example, phenol reacts with aqueous NaOH as follows.

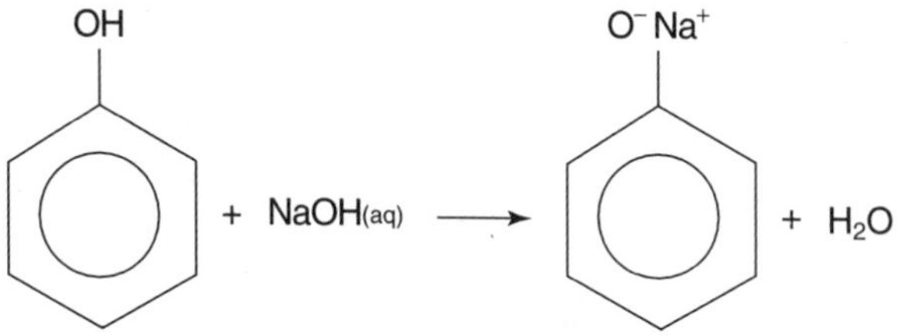

This is a typical neutralization reaction.

Because of their high acidity, phenols are often called **carbolic acids.** The phenol molecule is highly acidic because it has a partial positive charge on the oxygen atom due to resonance, and the anion that is formed by loss of a hydrogen ion is also resonance stabilized.

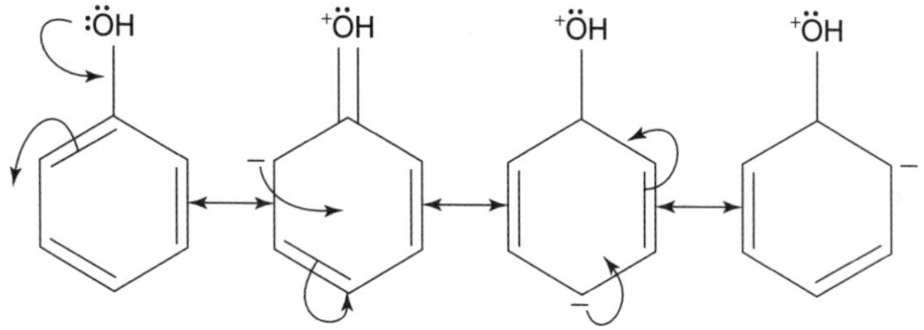

Resonance structures of phenol

Notice that three of the four contributing structures possess a positive charge on the oxygen atom of the molecule. Thus, the true hybrid structure must possess a partial positive charge. Because oxygen is an electronegative element, the electrons in the oxygen-hydrogen bond orbital are attracted to the oxygen atom, resulting in a partially positive hydrogen.

Loss of a hydrogen ion to a base creates a phenoxide ion that is resonance stabilized.

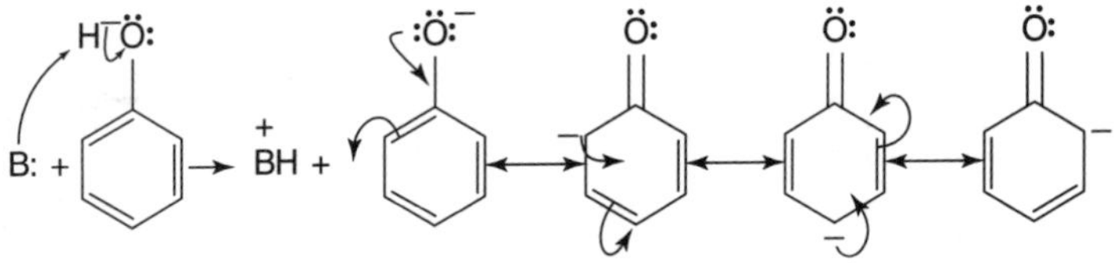

Notice that upon removal of the hydroxy hydrogen by a base, the phenoxide anion results. This anion is resonance stabilized by delocalization of an electron pair throughout the molecule, as shown by the contributing structures.

Synthesis of Phenols

You can prepare phenols in large quantities by the pyrolysis of the sodium salt of benzene sulfonic acid, by the Dow process, and by the air oxidation of cumene. Each of these processes is described below. You can also prepare small amounts of phenol by the peroxide oxidation of phenylboronic acid and the hydrolysis of diazonium salts.

Pyrolysis of sodium benzene sulfonate

In this process, benzene sulfonic acid is reacted with aqueous sodium hydroxide. The resulting salt is mixed with solid sodium hydroxide and fused at a high temperature. The product of this reaction is sodium phenoxide, which is acidified with aqueous acid to yield phenol.

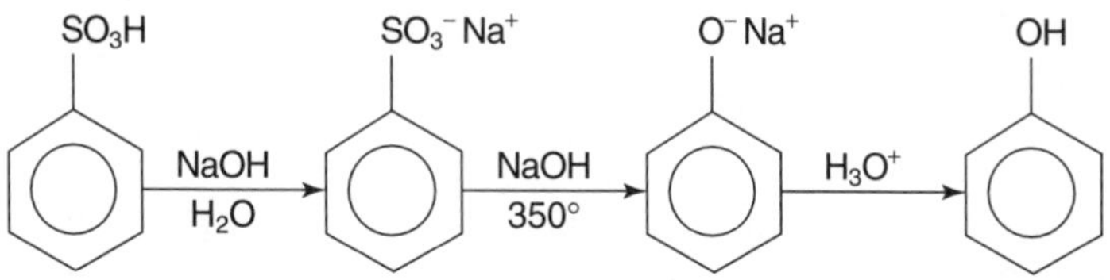

Dow process

In the Dow process, chlorobenzene is reacted with dilute sodium hydroxide at 300°C and 3000 psi pressure. The following figure illustrates the Dow process.

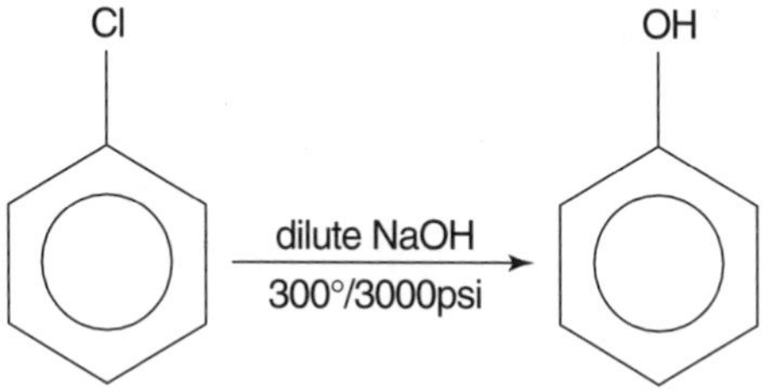

Air oxidation of cumene

The air oxidation of cumene (isopropyl benzene) leads to the production of both phenol and acetone, as shown in the following figure. The mechanisms for the formation and degradation of cumene hydroperoxide require closer looks, which are provided following the figure.

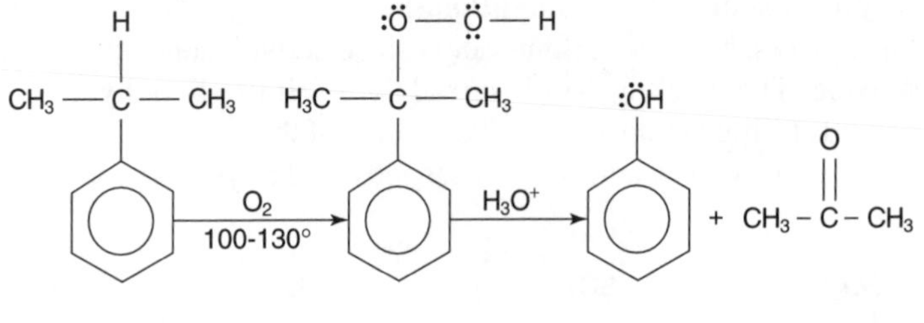

cumene hydroperoxide

Cumene hydroperoxide formation. The formation of the hydroperoxide proceeds by a free radical chain reaction. A radical initiator abstracts a hydrogen-free radical from the molecule, creating a tertiary free radical. The creation of the tertiary free radical is the initial step in the reaction.

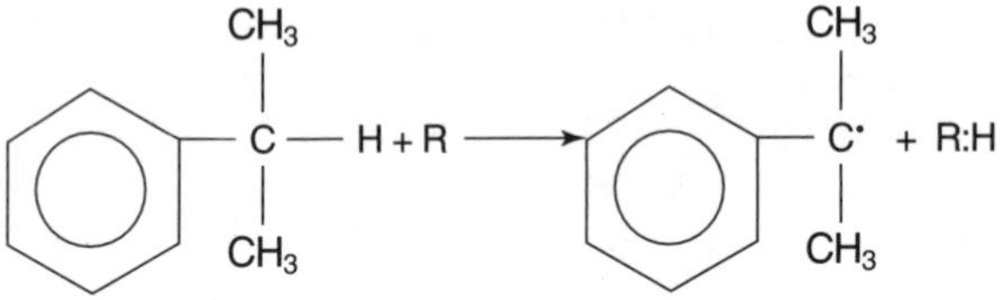

In the next step, the free radical is attracted to an oxygen molecule. This attraction produces the hydroperoxide free radical.

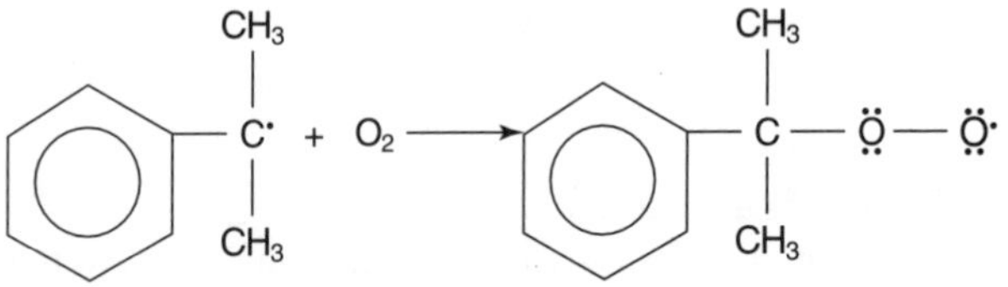

Finally, the hydroperoxide free radical abstracts a hydrogen free radical from a second molecule of cumene to form cumene hydroperoxide and a new tertiary free radical.

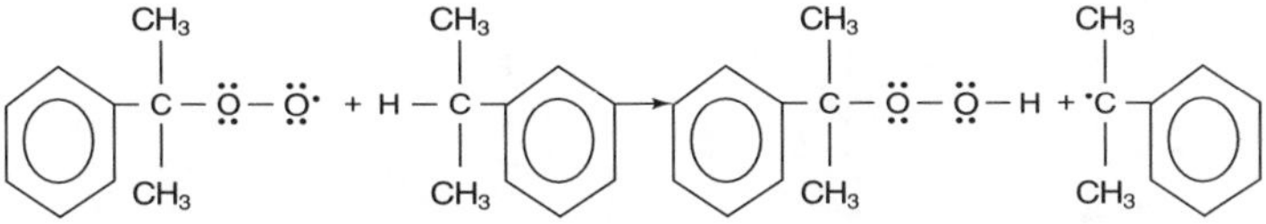

Cumene hydroperoxide degradation. The degradation of the cumene hydroperoxide proceeds via a carbocation mechanism. In the first step, a pair of electrons on the oxygen of the hydroperoxide's "hydroxyl group" is attracted to a proton of the H_3O^+ molecule, forming an oxonium ion.

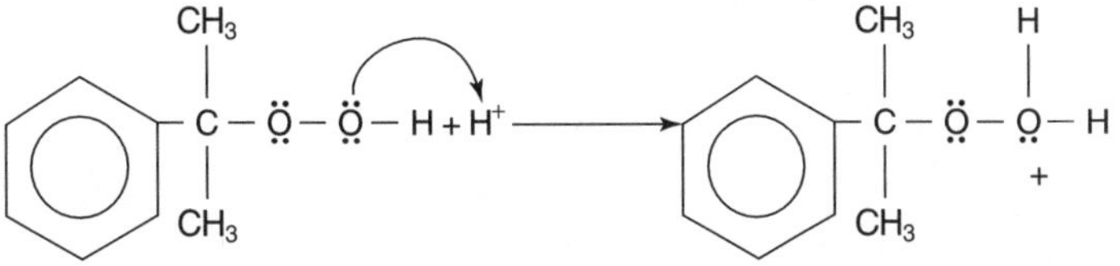

Next, the oxonium ion becomes stabilized when the positively charged oxygen leaves in a water molecule. This loss of a water molecule produces a new oxonium ion.

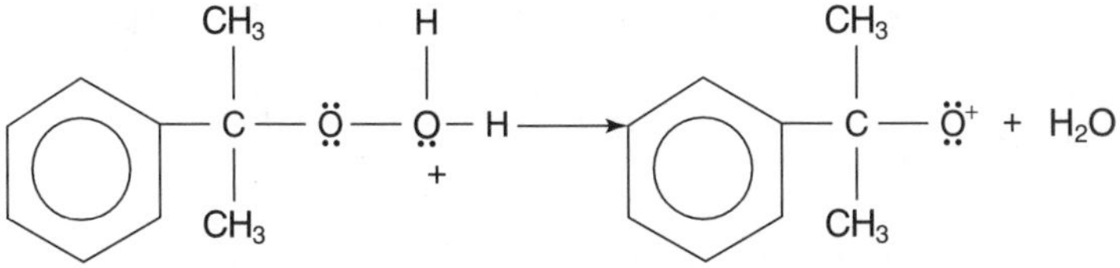

A phenide ion shift to the oxygen atom (which creates a tertiary carbocation) stabilizes the positively charged oxygen. (A phenide ion is a phenyl group with an electron bonding pair available to form a new bond to the ring.)

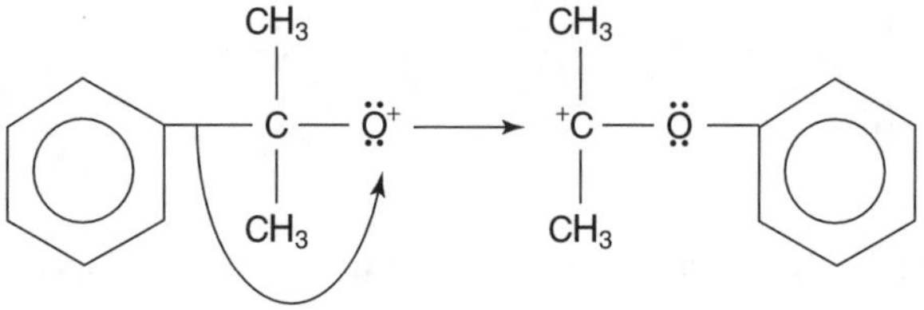

The carbocation is stabilized by an acid-base reaction with a water molecule, leading to the formation of an oxonium ion.

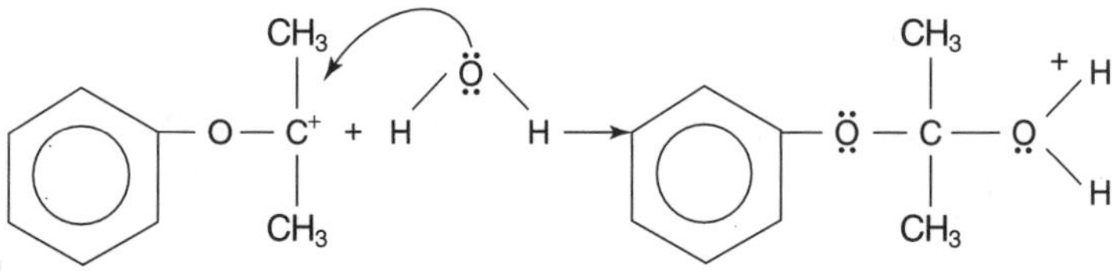

The loss of a proton stabilizes the oxonium ion.

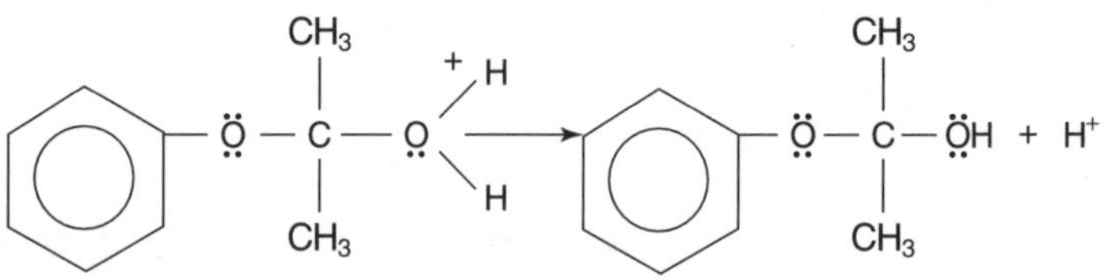

Next, a proton is picked up by the ether oxygen in an acid-base reaction, yielding a new oxonium ion.

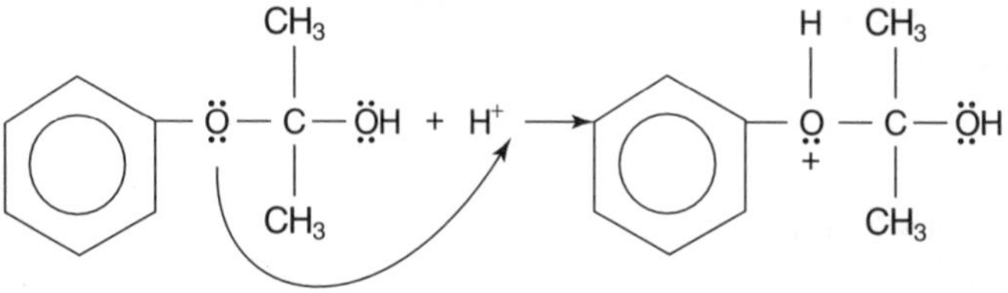

The positively charged ether oxygen pulls the electrons in the oxygen-carbon bond toward itself, thus delocalizing the charge over both of the atoms. The partial positive charge on the carbon attracts the nonbonding electron pair from the oxygen of the OH group, allowing the electrons in the original oxygen-carbon bond to be released back to the more electronegative oxygen atom.

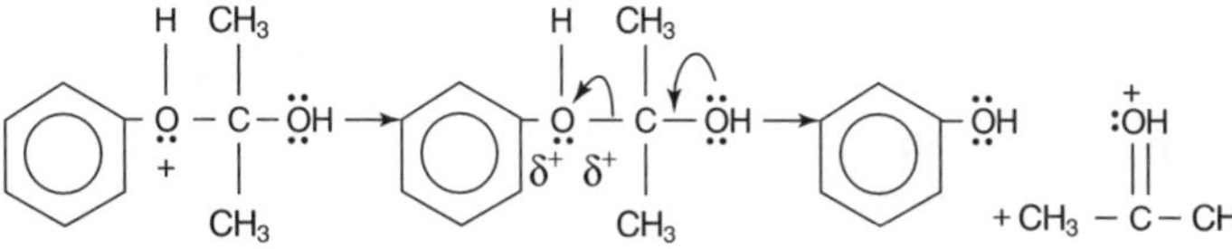

Finally, a proton is lost from the protonated acetone molecule, leading to the formation of acetone.

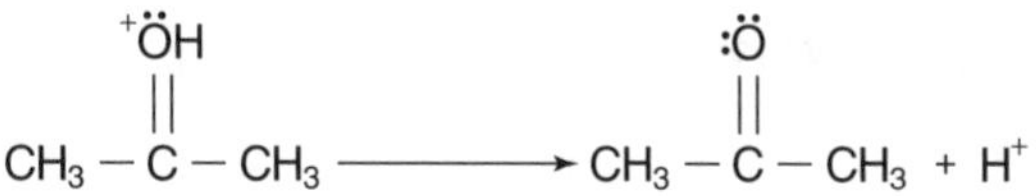

Reactions of Phenolic Hydrogen

As noted earlier in this chapter, phenols are acidic because of the ease with which the oxygen atom will release the hydrogen bonded to it.

This section describes typical reactions that occur as a result of the acidity of phenols.

Reactions with bases

Because phenol is acidic, it reacts with bases to form salts.

phenol sodium phenoxide

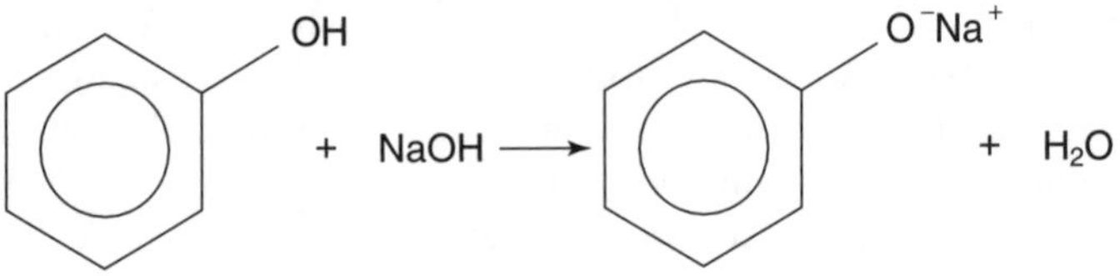

Esterification of phenol

Phenols form esters with acid anhydrides and acid chlorides.

phenol acetic anhydride phenylacetate

phenol acetyl chloride phenylacetate chloride ion

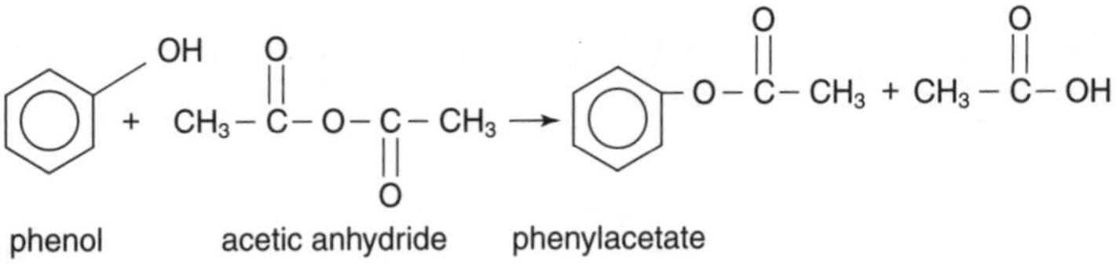

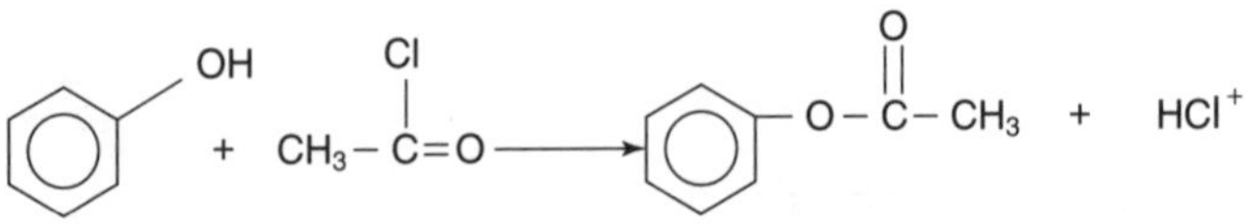

Williamson ether synthesis

Ethers are produced from phenol by the Williamson method via an S_N mechanism.

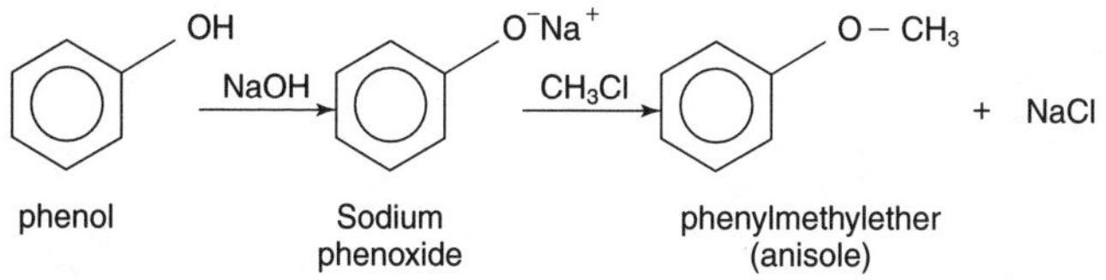

phenol   Sodium   phenylmethylether
phenoxide   (anisole)

Reactions of Phenolic Benzene Rings

The hydroxy group in a phenol molecule exhibits a strong activating effect on the benzene ring because it provides a ready source of electron density for the ring. This directing influence is so strong that you can often accomplish substitutions on phenols without the use of a catalyst.

Halogenation

Phenols react with halogens to yield mono-, di-, or tri-substituted products, depending on reaction conditions. For example, an aqueous bromine solution brominates all ortho and para positions on the ring.

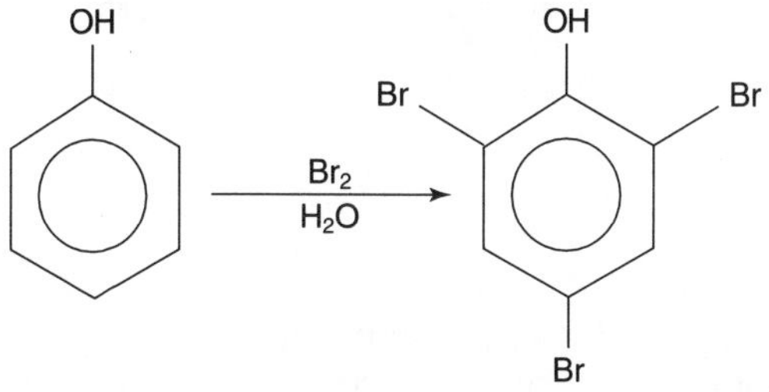

2,4,6-tribromophenol

Likewise, you can accomplish monobromination by running the reaction at extremely low temperatures in carbon disulfide solvent.

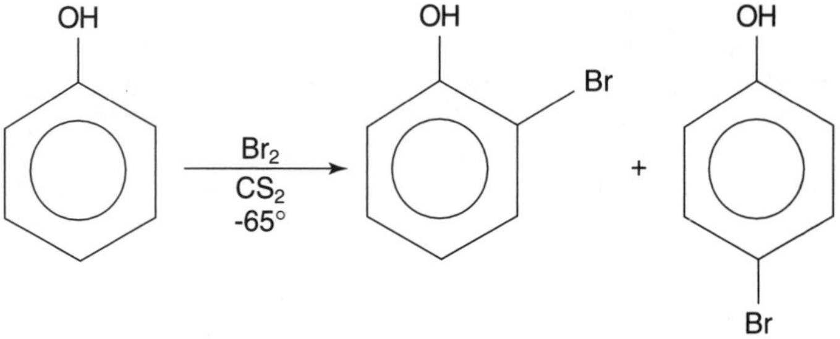

Nitration

Phenol, when treated with dilute nitric acid at room temperature, forms ortho- and para-nitrophenol.

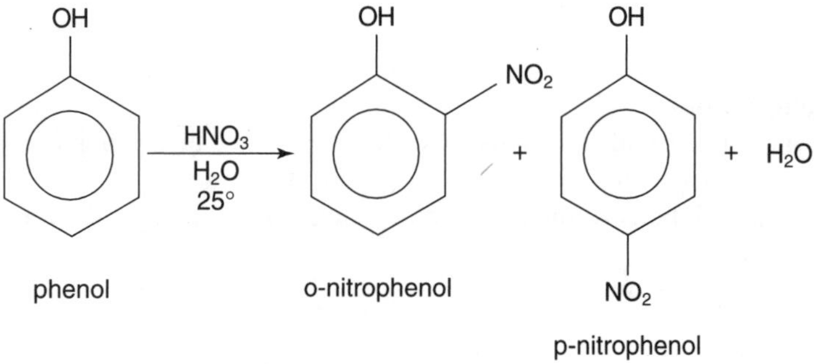

Sulfonation

The reaction of phenol with concentrated sulfuric acid is thermodynamically controlled. At 25°C, the ortho product predominates while at 100°C, the para product is the major product.

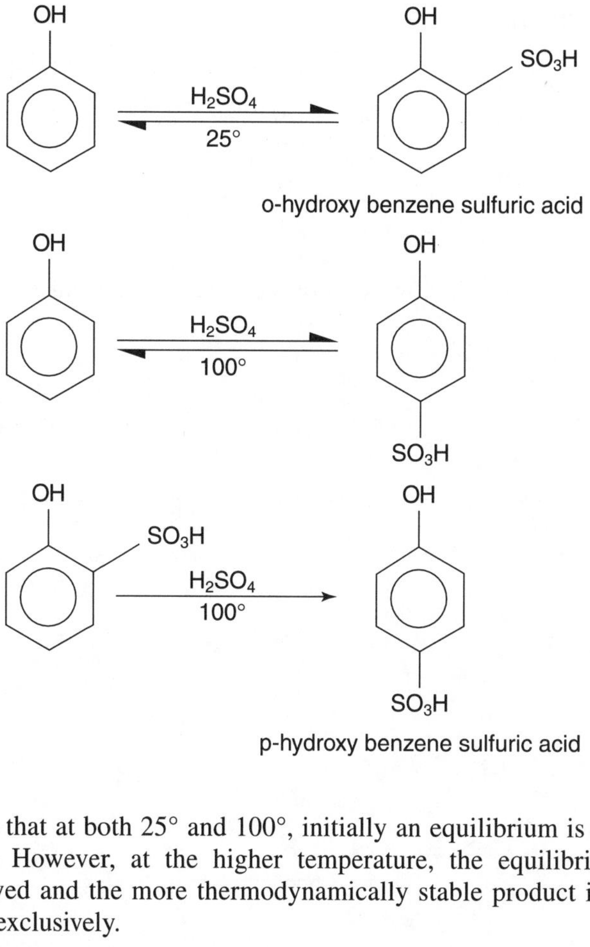

OH
H2SO4
25°
OH
SO3H
o-hydroxy benzene sulfuric acid
OH
H2SO4
100°
OH
SO3H
OH
SO3H
H2SO4
100°
OH
SO3H
p-hydroxy benzene sulfuric acid

Notice that at both 25° and 100°, initially an equilibrium is established. However, at the higher temperature, the equilibrium is destroyed and the more thermodynamically stable product is produced exclusively.

Kolbe reaction

The reaction of a phenoxide ion with carbon dioxide to produce a carboxylate salt is called the Kolbe reaction.

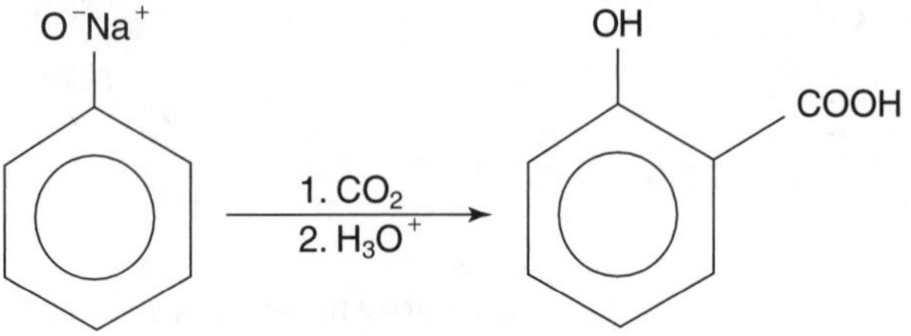

The Kolbe Reaction progresses via a carbanion intermediate.

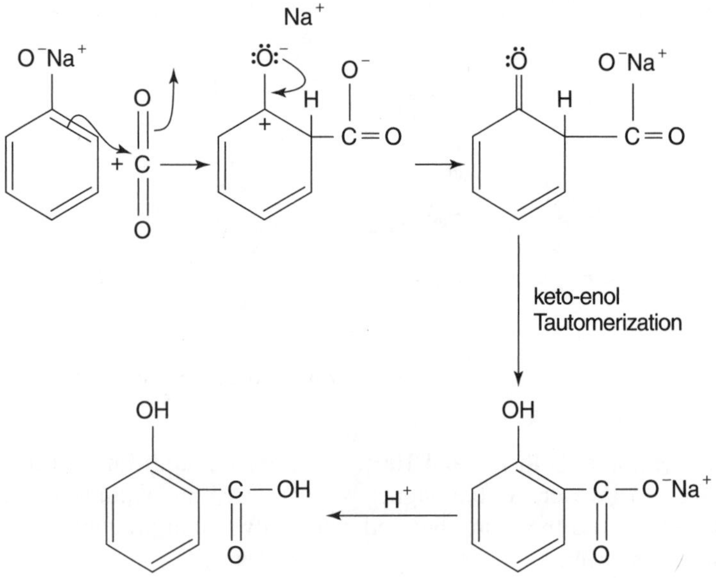

In this reaction, the electron deficient carbon atom in carbon dioxide is attracted to the electron rich π system of the phenol. The resulting compound undergoes keto-enol tautomerization to create the product.

Aryl Halides

An **aryl halide** is a compound formed by the substitution of a halogen atom for a hydrogen atom on benzene. Another name for an aryl halide is *halobenzene.*

Nomenclature

Aryl halides are named by prefixing the name of the halogen to benzene. For example:

chlorobenzene   bromobenzene   iodobenzene

Properties of aryl halides

The physical properties of unsubstituted aryl halides are much like those of the corresponding alkyl halides. Thus, boiling points, melting points, and solubilities of aryl halides are very similar to those of alkyl halides containing the same number of carbon atoms.

Synthesis of Aryl Halides

The two most common methods of preparing aryl halides are by direct halogenation of benzene and via diazonium salt reactions.

Halogenation of benzene

The halogenation of benzene to synthesize aryl halides is the oldest method known.

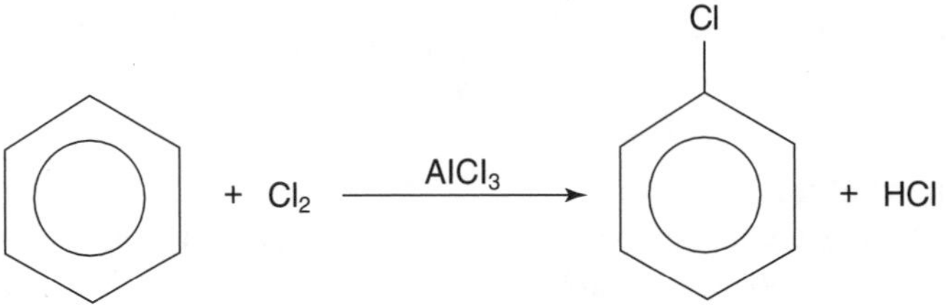

Sandmeyer reaction

A second method for preparing aryl halides is the Sandmeyer reaction. During a **Sandmeyer reaction,** a diazonium salt reacts with copper (I) bromide, copper (I) chloride, or potassium iodide to form the respective aryl halide. The diazonium salt is prepared from aniline by reaction with nitrous acid at cold temperatures.

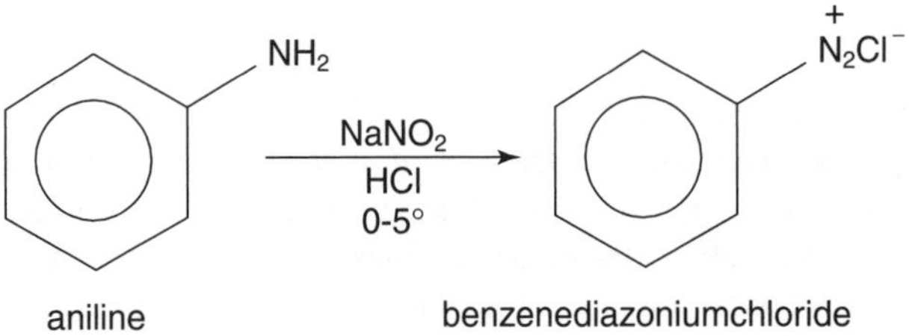

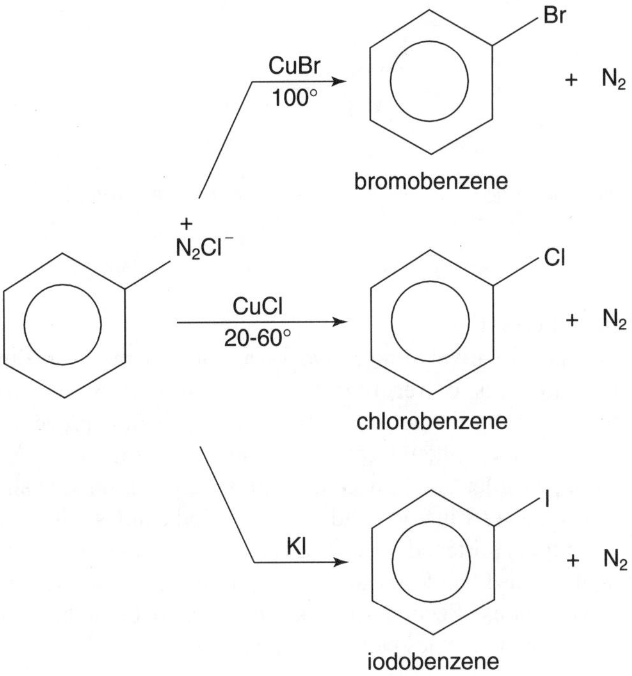
Br
CuBr
100°
+ N2
bromobenzene
+
N2Cl−
Cl
CuCl
20-60°
+ N2
chlorobenzene
I
KI
+ N2
iodobenzene

Reactions of Aryl Halides

Following are some typical reactions of aryl halides.

Grignard reaction

Aryl halides form Grignard reagents when reacted with magnesium.

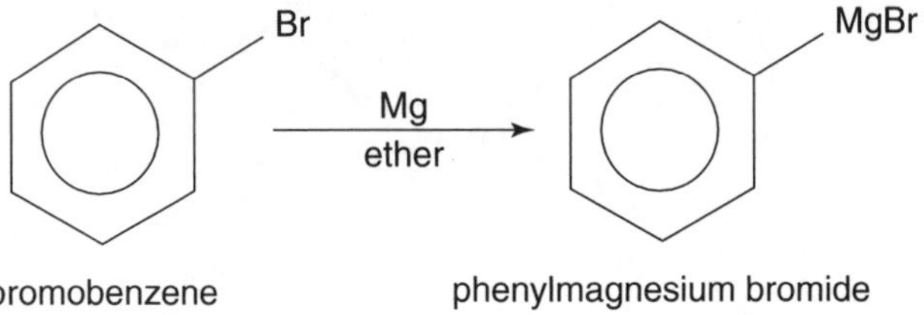

bromobenzene   phenylmagnesium bromide

Substitution reaction

Aryl halides are relatively unreactive toward nucleophilic substitution reactions. This lack of reactivity is due to several factors. Steric hindrance caused by the benzene ring of the aryl halide prevents S_N2 reactions. Likewise, phenyl cations are unstable, thus making S_N1 reactions impossible. In addition, the carbon-halogen bond is shorter and therefore stronger in aryl halides than in alkyl halides. The carbon-halogen bond is shortened in aryl halides for two reasons. First, the carbon atom in aryl halides is sp^2 hybridized instead of sp^3 hybridized as in alkyl halides. Second, the carbon-halogen bond has partial double bond characteristics because of resonance.

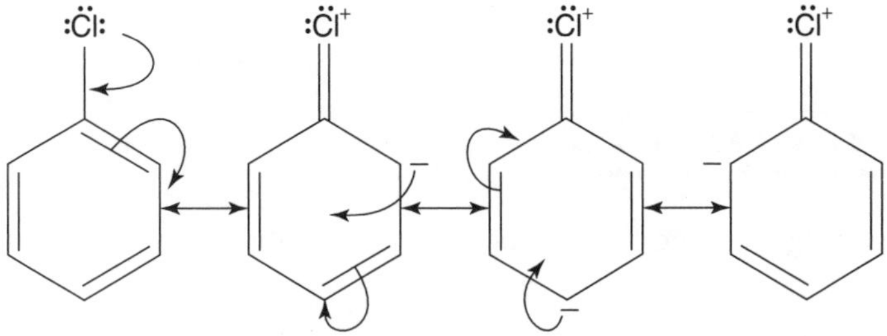

Because three of the four resonance structures show a double bond between the carbon and halogen atoms, the hybrid structure must have double bond character.

S$_N$AR reactions

Nucleophilic substitution reactions can occur with aryl halides, provided that strong electron-withdrawing groups (deactivators) are located ortho and/or para to the carbon atom that's attached to the halogen. (This arrangement makes the carbon susceptible to nucleophilic attack.)

The examples below illustrate S$_N$ substitutions on deactivated aryl halides.

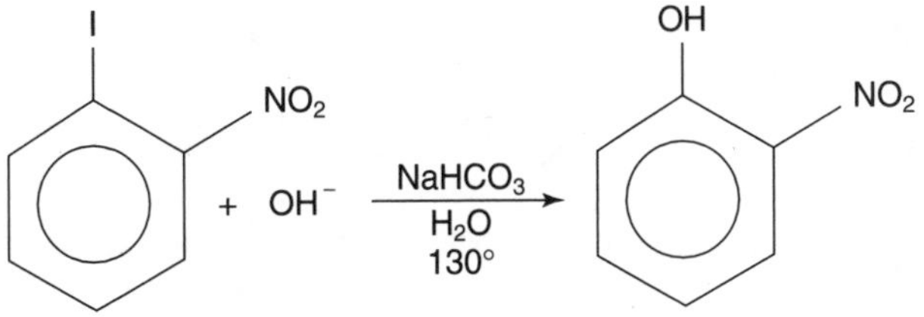

o-nitroiodobenzene   o-nitrophenol

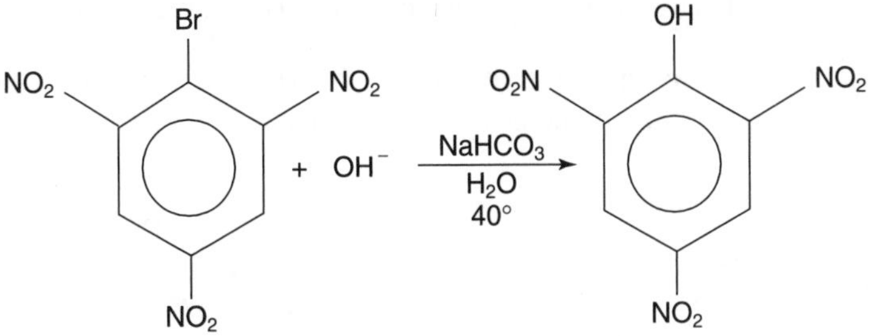

2,4,6-trinitrobromobenzene   2,4,6-trinitrophenol

S$_N$AR mechanism

The **S$_N$AR mechanism** is an addition-elimination mechanism that proceeds through a carbanion with delocalized electrons (a Meisenheimer complex). The following steps show the mechanism for the formation of p-nitrophenol from p-nitroiodobenzene.

1. The nitro group, a strong deactivating group, produces a partial positive charge on the carbon that bears the halogen atom in the aryl halide.

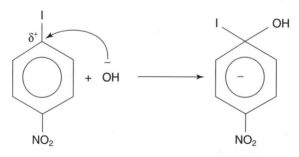

Because one of the resonance structures has a positive charge on the carbon attached to the halogen, this carbon acts as a weak nucleus.

2. The hydroxide ion is attracted to the weak nucleus, forming a carbocation with delocalized electrons.

carbanion
(Meisenheimer Complex)

3. The complex eliminates an iodide ion to form a phenol.

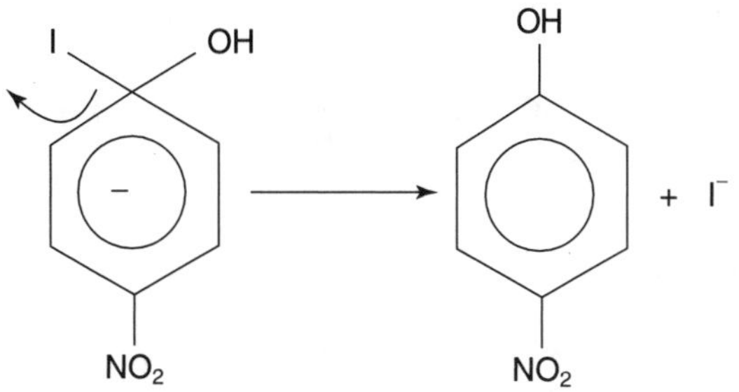

Elimination-addition reactions

As noted earlier in this section, aryl halides generally do not undergo substitution reactions. However, under conditions of high temperature and pressure, these compounds can be forced to undergo substitution reactions. For example, under high temperature and pressure, chlorobenzene can be converted into sodium phenoxide when reacted with sodium hydroxide.

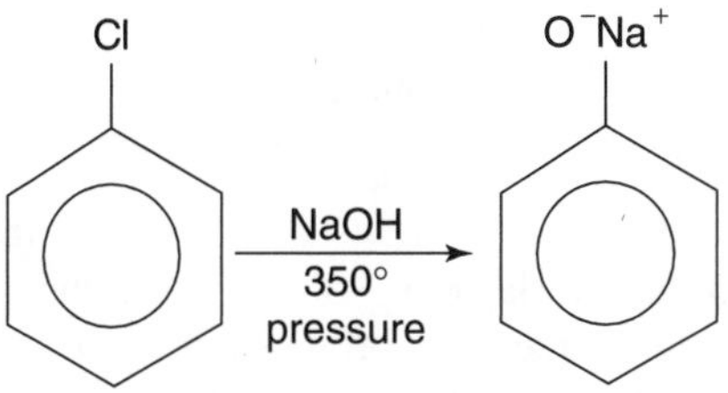

Similarly, at a very low temperature, bromobenzene reacts with potassium amide (KNH_2) dissolved in liquid ammonia to form aniline.

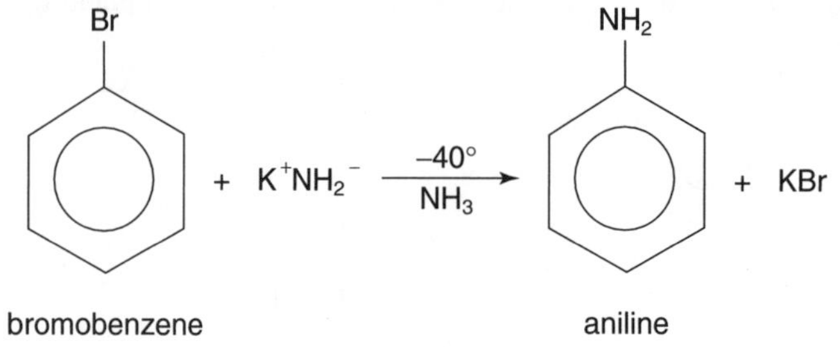

Br   + $K^+NH_2^-$ $\xrightarrow[\text{NH}_3]{-40°}$ NH$_2$ +   KBr

bromobenzene   aniline

Mechanism for aniline formation

The elimination-addition mechanism for the formation of aniline proceeds via a benzyne intermediate. A **benzyne** is a benzene molecule that contains a theoretical triple bond. Thus, the following structure represents benzyne:

benzyne

A triple bond doesn't exist in the true benzyne structure. The extra bond results from the overlap of sp^2 orbitals on adjacent carbon atoms of the ring. The axes of these sp^2 orbitals are in the same plane as the ring, and therefore, they don't overlap with the π orbitals of the aromatic system. Consequently, there's little to no interference with the aromatic system. The additional bond is weak and benzyne is thus highly unstable and highly reactive. Figure 4-1 shows the true structure of benzyne:

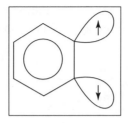

Figure 4-1

The following four steps outline the mechanism for aniline formation.

1. An amide ion, a very strong base, removes a weak proton from the carbon that is alpha to the carbon bonded to the bromine.

2. The carbanion electrons are stabilized by being attracted to the electronegative bromine, which results with the loss of a bromide ion.

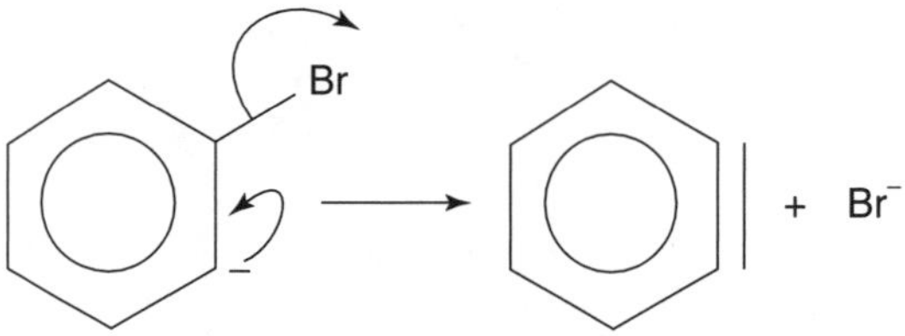

benzyne

3. The highly unstable and very reactive benzyne reacts with a second amide ion, creating a new carbanion.

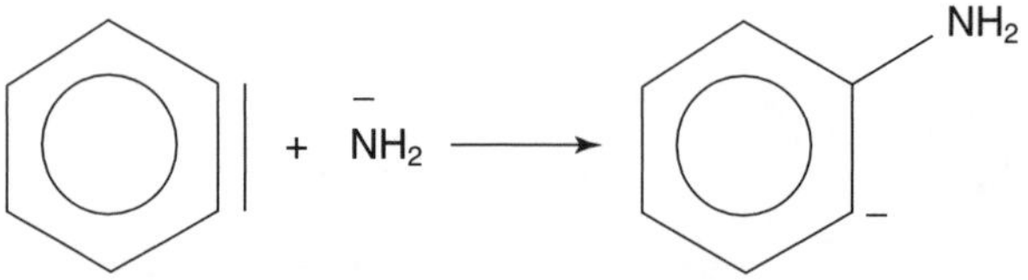

4. The new carbanion abstracts a proton from an ammonia molecule in an acid-base reaction, leading to the formation of aniline.

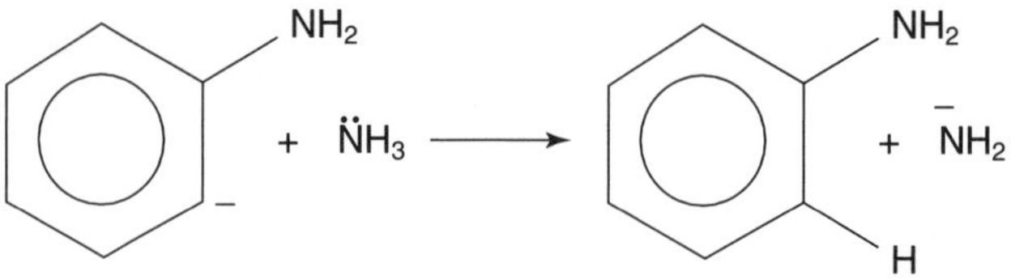

Alcohols

Alcohols are hydroxy-substituted alkanes, alkenes, or alkynes in which the substitution occurs on a saturated carbon.

The general formula for alcohols is R — OH, where the R group can represent the alkyl, alkenyl, or alkynal groups. In the case of substitution on alkenes and alkynes, only saturated carbons may be substituted. For example, the following compounds are all alcohols:

$CH_3 - CH_2 - CH_2 - OH$ $H_2C = CH - CH_2 - OH$ $HC \equiv C - CH_2 - OH$

   propanol   2–propenol   2–propynol

If the hydroxyl group were substituted for a hydrogen on an unsaturated carbon, an alcohol would not form. For example, substituting the hydroxyl group for a terminal hydrogen of 1-propene gives an unstable enol that tautomerizes to a ketone.

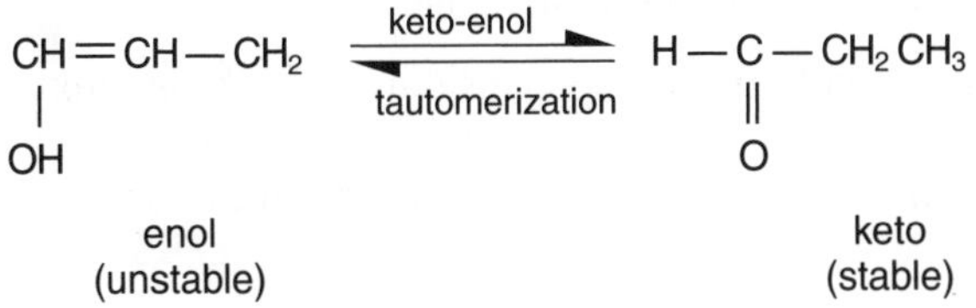

$$CH = CH - CH_2 \quad \xrightleftharpoons[\text{tautomerization}]{\text{keto-enol}} \quad H - \underset{O}{\overset{||}{C}} - CH_2 CH_3$$
$$|$$
$$OH$$

                enol   keto
         (unstable)   (stable)

Nomenclature

You can use both the common and IUPAC systems to name alcohols. In the common system, you name an alcohol by listing the alkyl group and adding the word *alcohol*. Following are some examples of alcohols and their common names:

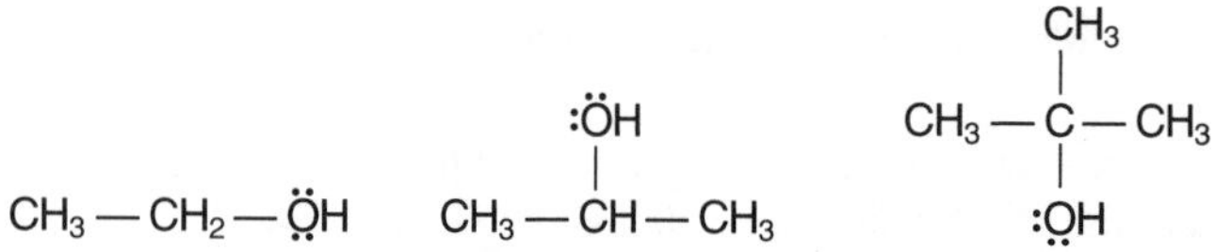

$$CH_3-CH_2-\overset{..}{\underset{..}{O}}H \qquad CH_3-\overset{\overset{:\ddot{O}H}{|}}{C}H-CH_3 \qquad CH_3-\overset{\overset{CH_3}{|}}{\underset{\underset{:\ddot{O}H}{|}}{C}}-CH_3$$

ethyl alcohol   isopropyl alcohol   t-butyl alcohol

In the IUPAC system, use the following series of rules to name alcohols:

1. Pick out the longest continuous chain to which the hydroxyl group is directly attached. The parent name of the alcohol comes from the alkane name for the same chain length. Drop the -e ending and add -ol.

2. Number the parent chain so that the carbon bearing the hydroxyl group has the lowest possible number. Place the number in front of the parent name.

3. Locate and name substituents other than the hydroxyl group.

The following examples show how you apply these rules:

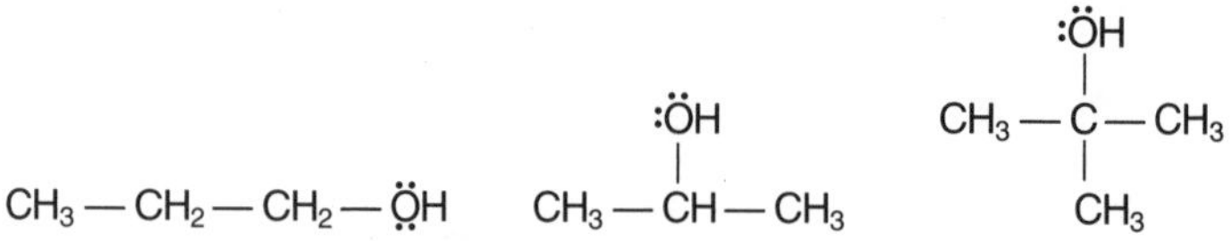

$$CH_3-CH_2-CH_2-\overset{..}{\underset{..}{O}}H \qquad CH_3-\overset{\overset{:\ddot{O}H}{|}}{C}H-CH_3 \qquad CH_3-\overset{\overset{:\ddot{O}H}{|}}{\underset{\underset{CH_3}{|}}{C}}-CH_3$$

1-propanol   2-propanol   2-methyl-2-propanol

You may classify alcohols as primary (1°), secondary (2°), or tertiary (3°), based on the class of carbon to which the hydroxyl group (—OH) is directly bonded. For example, 1-propanol is a 1° alcohol, 2-propanol is a 2° alcohol, and 2-methyl-2-propanol is a 3° alcohol.

Physical properties
Alcohols contain both a polar —OH group and a nonpolar alkyl group. As a result of this composition, alcohols that have small alkyl chains tend to be water soluble. As alkyl chain length increases, water solubility decreases.

Through the OH group, alcohols are capable of forming hydrogen bonds to themselves, other alcohols, neutral molecules, and anions. This bond formation leads to abnormally high boiling points compared to other organic molecules of similar carbon chain length.

Synthesis of Alcohols

Alcohols can be prepared by the hydration of alkenes or by the reduction of aldehydes, ketones, acids, and esters.

Hydration of alkenes
The elements of water can be added to the double-bonded carbons of an alkene in either a Markovnikov's or an anti-Markovnikov's manner. As shown in the following figure, a hydrogen ion catalyzes the Markovnikov's addition.

$$CH_3 - CH = CH_2 + H_2O \xrightarrow{H^+} CH_3 - \overset{\overset{\displaystyle OH}{|}}{CH} - CH_3 + CH_3 - CH_2 - CH_2 - OH$$

propene   2-propanol   1-propanol

(major product)   (minor product)

The anti-Markovnikov's addition results from a hydroboration-oxidation reaction.

$$CH_3 - CH = CH_2 + (CH_3)_2 \, BH \xrightarrow[-OH]{H_2O_2} CH_3 \, CH_2 \, CH_2 \, OH + CH_3 - \overset{\overset{\displaystyle \quad}{|}}{CH} - CH_3$$

(major product)   OH

(minor product)

You can find the mechanisms for both the Markovnikov's and anti-Markovnikov's addition of water in *CliffsQuickReview Organic Chemistry I.*

Reduction of aldehydes and ketones

An aldehyde has a structural formula of $R - \overset{\overset{\displaystyle O}{\|}}{C} - H$

while the structural formula of a ketone is $R - \overset{\overset{\displaystyle O}{\|}}{C} - R'$

In these formulas, the R or R´ group may be either an aliphatic or aromatic group. In a ketone, the R and R´ groups may represent the same group or different groups. These types of compounds are best reduced by complex metal hydrides, such as lithium aluminum hydride (LiAlH₄) or sodium borohydride (NaBH₄).

Following are two examples of complex metal reductions:

$$CH_3 - CH_2 - CH_2 - CH_2 \overset{\overset{\displaystyle O}{\|}}{C} - H \quad \xrightarrow[\text{2. }H_2O]{\substack{\text{1. LiAlH}_4 \\ \text{Ether}}} \quad CH_3 - CH_2 - CH_2 - CH_2 - OH$$

butanol   1-butanol

$$CH_3 - CH_2 - \overset{\overset{\displaystyle O}{\|}}{C} - CH_3 \quad \xrightarrow[\text{H}_2O]{\text{NaBH}_4} \quad CH_3 - CH_2 - \overset{\overset{\displaystyle OH}{|}}{CH} - CH_3$$

2-butanone   2-butanol

Lithium aluminum hydride is a very strong reducing agent that will reduce many functional groups in addition to aldehydes and ketones. Sodium borohydride is a much weaker reducing agent that basically will reduce only aldehydes and ketones to alcohols.

You can also catalytically reduce aldehydes and ketones to produce 1° and 2° alcohols. Reduction conditions are very similar to those used to reduce alkene double bonds. If a molecule possesses both a double bond and an aldehyde or ketone functional group, reduction of the aldehyde or ketone group is best carried out using sodium borohydride. The reduction of cyclohexanone by hydrogen gas with a platinum catalyst produces cyclohexanol in good yield.

cyclohexanone   cyclohexanol

Reduction of carboxylic acids

The reduction of a carboxylic acid:

$$
\begin{array}{c}
O \\
\parallel \\
R - C - OH
\end{array}
$$

leads to the formation of a primary alcohol:

$$
\begin{array}{c}
O \\
\parallel \\
R - C - OH
\end{array}
\xrightarrow{\text{H}_2}
R - CH_2OH
$$

This reduction requires a very strong reducing agent, and lithium aluminum hydride is the standard choice.

$$
\begin{array}{c}
O \\
\parallel \\
CH_3 - C - OH
\end{array}
\xrightarrow[\substack{\text{Ether} \\ \\ 2.\ \text{H}_2\text{O}}]{1.\ \text{LiAlH}_4}
CH_3 - CH_2 - OH
$$

acetic acid   ethanol

Diborane, B_2H_6, also reduces carboxylic acids to alcohols.

$$
\begin{array}{c}
O \\
\parallel \\
CH_3 - CH_2 - C - OH
\end{array}
\xrightarrow[\substack{\text{THF} \\ \\ 2.\ \text{H}^+}]{1.\ \text{B}_2\text{H}_6}
CH_3 - CH_2 - CH_2 - OH
$$

propionic acid   propanol

Catalytic hydrogenation gives very poor yields and is not usually used for this type of reaction.

Reduction of esters

Esters, like carboxylic acids, are normally reduced with lithium aluminum hydride. In these reactions, two alcohols are formed. An example is the reduction of methyl benzoate to benzyl alcohol and methanol.

methyl benzoate   benzyl alcohol

Grignard reaction with aldehydes and ketones

The Grignard reaction is the only simple method available that is capable of producing primary, secondary, and tertiary alcohols. To produce a primary alcohol, the Grignard reagent is reacted with formaldehyde.

$$CH_3MgBr \quad + \quad H-\overset{\overset{H}{|}}{C}=O \quad \xrightarrow{HCl} \quad CH_3-CH_2-OH + Mg(Br)Cl$$

methyl magnesium   formaldehyde   ethanol
bromide   (1° alcohol)
(Grignard reagent)

Reacting a Grignard reagent with any other aldehyde will lead to a secondary alcohol.

$$CH_3MgBr \quad + \quad CH_3-\overset{\overset{O}{\|}}{CH} \quad \longrightarrow \quad CH_3-\overset{\overset{OH}{|}}{CH}-CH_3 + Mg(Br)Cl$$

acetaldehyde   2-propanol
(2° alcohol)

Finally, reacting a Grignard reagent with a ketone will generate a tertiary alcohol.

$$CH_3MgBr \quad + \quad CH_3-\overset{\overset{\displaystyle O}{\|}}{C}-CH_3 \quad \longrightarrow \quad CH_3-\overset{\overset{\displaystyle CH_3}{|}}{\underset{\underset{\displaystyle OH}{|}}{C}}-CH_3 \; + \; Mg(Br)Cl$$

acetone

t-Butyl alcohol

(3° alcohol)

Reactions of Alcohols

Alcohols are capable of being converted to metal salts, alkyl halides, esters, aldehydes, ketones, and carboxylic acids.

Metal salt formation

Alcohols are only slightly weaker acids than water, with a K_a value of approximately 1×10^{-16}. The reaction of ethanol with sodium metal (a base) produces sodium ethoxide and hydrogen gas.

$$CH_3-CH_2-OH \; + \; Na \; \longrightarrow \; CH_3-CH_2-O^-Na^+ \; + \; H_2$$

This reaction is identical to the reaction of sodium metal with water.

$$HOH \; + \; Na \; \longrightarrow \; NaOH \; + \; H_2$$

However, the latter reaction occurs faster because of the increased acidity of water (K_a value of 1×10^{-15}). Likewise, similar reactions occur with potassium metal.

The acidity of alcohols decreases while going from primary to secondary to tertiary. This decrease in acidity is due to two factors: an increase of electron density on the oxygen atom of the more highly-substituted alcohol, and steric hindrance (because of the alkyl groups, which inhibit solvation of the resulting alkoxide ion). Both of these situations increase the activation energy for proton removal.

The basicity of alkoxide ions increases while going from primary to tertiary. This increase in basicity occurs because the conjugate base of a weak acid is strong. The weaker the acid, the stronger the conjugate base.

Alkyl halide formation

Alcohols are converted to alkyl halides by S_N1 and S_N2 reactions with halogen acids.

$$CH_3CH_2OH \ + \ HCl \ \xrightarrow{\text{heat}} \ CH_3 - CH_2 - Cl \ + \ H_2O \quad (S_N2)$$

$$\underset{\overset{|}{OH}}{\overset{\overset{CH_3}{|}}{H_3 - C - CH_3}} \ + \ HCl \ \xrightarrow{\text{heat}} \ \underset{\overset{|}{Cl}}{\overset{\overset{CH_3}{|}}{CH_3 - C - CH_3}} \ + \ H_2O \quad (S_N1)$$

Primary alcohols favor S_N2 substitutions while S_N1 substitutions occur mainly with tertiary alcohols.

A more efficient method of preparing alkyl halides from alcohols involves reactions with thionyl chloride ($SOCl_2$).

$$CH_3 - CH_2 - OH \ + \ SOCl_2 \ \xrightarrow{\text{heat}} \ CH_3 - CH_2 - Cl \ + \ SO_2 \ + \ HCl$$

This reaction is rapid and produces few side reaction products. In addition, the sulfur dioxide and hydrogen chloride formed as byproducts are gasses and therefore easily removed from the reaction. Mechanistically, the alcohol initially reacts to form an inorganic ester.

$$\text{R CH}_2\overset{\cdot\cdot}{\underset{\cdot\cdot}{O}}\text{H} + \text{Cl}-\overset{\overset{\cdot\cdot}{O}}{\underset{\|}{S}}-\text{Cl} \longrightarrow \text{R CH}_2\overset{\overset{O}{\|}}{O}\text{SCl} + \text{H}^+ + :\overset{\cdot\cdot}{\underset{\cdot\cdot}{Cl}}:^-$$

The chloride ion produced by this reaction, acting as a nucleophile, attacks the ester in an S_N2 fashion to yield molecules of sulfur dioxide, hydrogen chloride, and an alkyl halide.

$$\text{H}^+ + :\overset{\cdot\cdot}{\underset{\cdot\cdot}{Cl}}:^- + \underset{\underset{R}{|}}{\text{CH}_2}-\overset{\cdot\cdot}{\underset{\cdot\cdot}{O}}-\overset{\overset{\cdot\cdot}{O}}{\underset{\|}{S}}\overset{\cdot\cdot}{\underset{\cdot\cdot}{Cl}}: \longrightarrow \text{ClCH}_2\text{R} + :\overset{\cdot\cdot}{O}=S=\overset{\cdot\cdot}{O}: + \text{H}\overset{\cdot\cdot}{\underset{\cdot\cdot}{Cl}}:$$

Because the reaction proceeds mainly by an S_N2 mechanism, the alkyl halide produced from an optically active alcohol will have the opposite relative configuration from the alcohol from which it was formed.

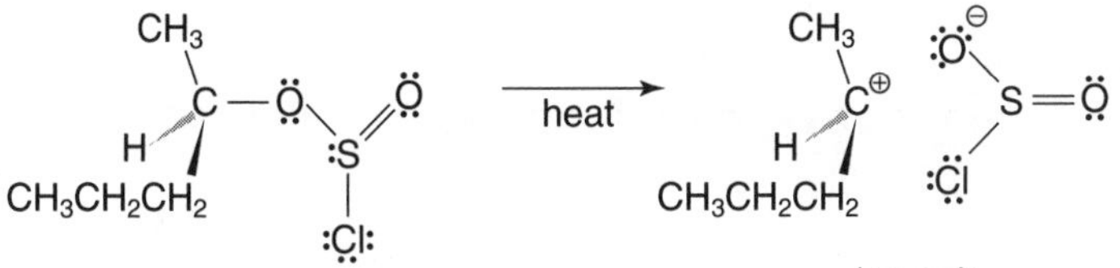

ion pair

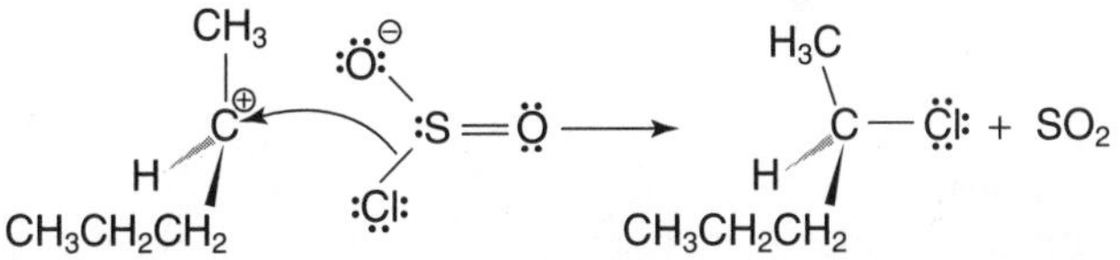

ion pair

Because thionyl bromide is relatively unstable, alkyl bromides are normally prepared by reacting the alcohol with phosphorous tribromide (PBr₃).

$$CH_3 — OH + PBr_3 \longrightarrow CH_3Br + H_3PO_3$$

This reaction proceeds via a two-step mechanism. In the first step, the alcohol reacts with the phosphorous tribromide.

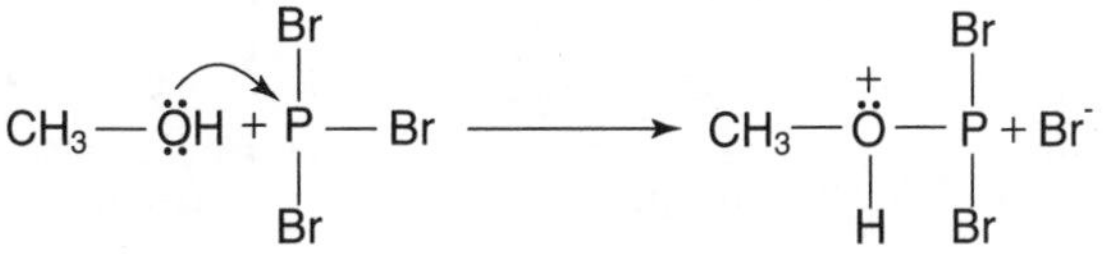

The second step is an S_N1 or S_N2 substitution in which the bromide ion displaces the dibromophosphorous group.

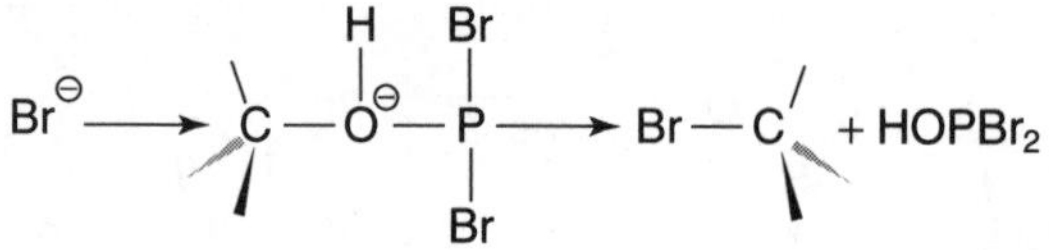

In a similar manner, alkyl iodides are prepared by reacting an alcohol with phosphorous triiodide.

Ester formation

Esters are compounds that are commonly formed by the reaction of oxygen-containing acids with alcohols. The ester functional group is

the $-\overset{\overset{\text{O}}{\|}}{\text{C}}-\text{OR}$ group.

Alcohols can be converted to esters by means of the Fischer Esterification Process. In this method, an alcohol is reacted with a carboxylic acid in the presence of an inorganic acid catalyst.

Because the reaction is an equilibrium reaction, in order to receive a good yield, one of the products must be removed as it forms. Doing this drives the equilibrium to the product side.

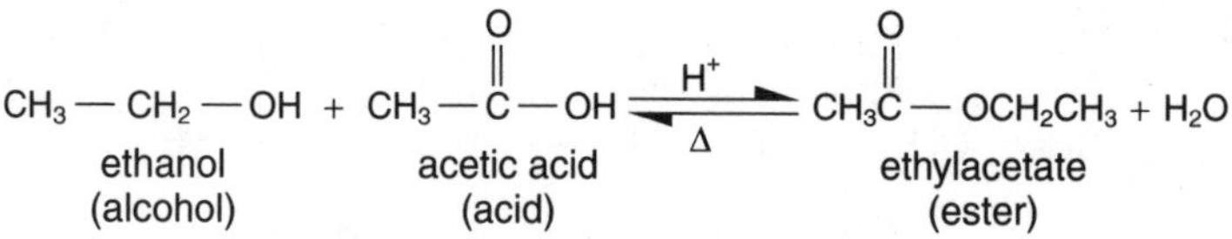

$$CH_3-CH_2-OH \ + \ CH_3-\overset{\overset{\text{O}}{\|}}{C}-OH \ \underset{\Delta}{\overset{H^+}{\rightleftharpoons}} \ CH_3\overset{\overset{\text{O}}{\|}}{C}-OCH_2CH_3 \ + \ H_2O$$

ethanol   acetic acid   ethylacetate
(alcohol)   (acid)   (ester)

The mechanism for this type of reaction takes place in seven steps:

1. The mechanism begins with the protonation of the acetic acid.

$$CH_3 - \overset{\overset{\displaystyle \ddot{O}:}{\|}}{C} - OH \; + \; H^+ \longrightarrow CH_3 - \overset{\overset{\displaystyle :\overset{+}{O}-H}{\|}}{C} - OH$$

2. The π electrons of the carboxyl group, $\overset{\displaystyle |}{\underset{\displaystyle |}{C}} = \ddot{O}:$, migrate to pick up the positive charge.

$$CH_3 - \overset{\overset{\displaystyle :\overset{+}{O}-H}{\|}}{C} - OH \longleftrightarrow CH_3 - \overset{\overset{\displaystyle :O-H}{\|}}{\underset{\displaystyle +}{C}} - OH$$

3. The oxygen of the alcohol molecule attacks the carbocation.

$$CH_3 - CH_2 - \overset{..}{\underset{..}{O}}H \; + \; CH_3 - \overset{\overset{\displaystyle :\ddot{O}H}{|}}{\underset{\displaystyle +}{C}} - \ddot{O}H \longrightarrow \begin{array}{c} CH_3 - \overset{\overset{\displaystyle :\ddot{O}H}{|}}{C} - \ddot{O}H \\ | \\ CH_3 - CH_2 - \overset{..}{\underset{\displaystyle +}{O}} - H \end{array}$$

4. The oxonium ion that forms loses a proton.

$$\begin{array}{c} \overset{\displaystyle OH}{|} \\ CH_3 - C - OH \\ | \\ CH_3CH_2 - \overset{..}{\underset{..}{O}} - H \end{array} \longrightarrow \begin{array}{c} \overset{\displaystyle OH}{|} \\ CH_3 - C - OH \\ | \\ CH_3 - CH_2 - \overset{..}{\underset{..}{O}}: \end{array} + H^+$$

5. One of the hydroxyl groups is protonated to form an oxonium ion.

$$CH_3-\underset{\underset{CH_3CH_2-O}{|}}{\overset{:\ddot{O}H}{\overset{|}{C}}}-\ddot{O}H \xrightarrow{H^+} CH_3-\underset{\underset{CH_3CH_2-O}{|}}{\overset{:\ddot{O}H}{\overset{|}{C}}}-\underset{+}{\overset{H}{\overset{|}{O}H}}$$

6. An unshared pair of electrons on another hydroxy group reestablishes the carbonyl group, with the loss of a water molecule.

$$CH_3-\underset{\underset{CH_3-CH_2-O}{|}}{\overset{:\ddot{O}H}{\overset{|}{C}}}-\underset{+}{\overset{H}{\overset{|}{O}H}} \longrightarrow CH_3-\underset{\underset{OCH_3CH_2}{||}}{\overset{\overset{+}{O}-H}{C}} + H_2O$$

7. The oxonium ion loses a proton, which leads to the production of the ester.

$$CH_3-\underset{\underset{OCH_3CH_2}{|}}{\overset{\overset{+}{:O}-H}{C}} \longrightarrow CH_3-\overset{:\ddot{O}}{\overset{||}{C}}-OCH_2CH_3 + H^+$$

Alkyl sulfonate formation. Alcohols may be converted to alkyl sulfonates, which are sulfonic acid esters. These esters are formed by reacting an alcohol with an appropriate sulfonic acid. For example, methyl tosylate, a typical sulfonate, is formed by reacting methyl alcohol with tosyl chloride.

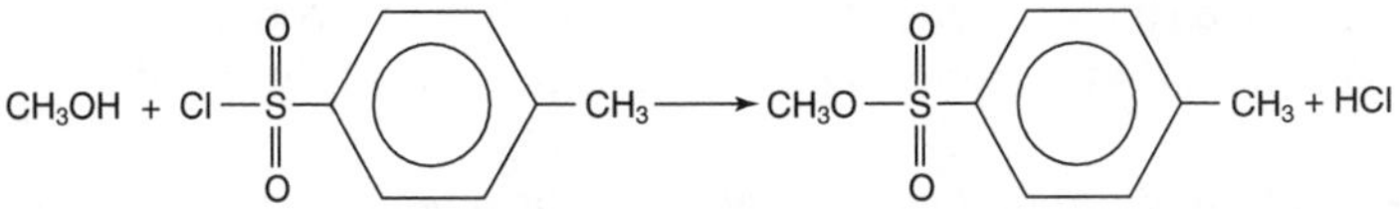

tosyl chloride
(p-toluene sulfonyl chloride)

methyl tosylate
(methyl-p-toluene sulfonate)

Other sulfonyl halides that form alkyl sulfonates include:

brosyl

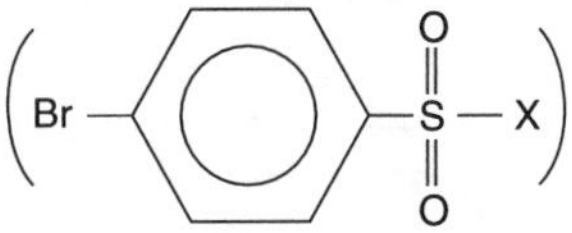

mesyl

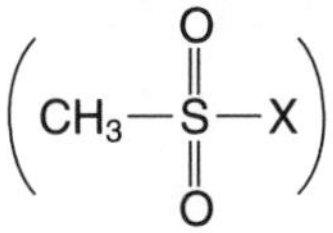

and trifyl

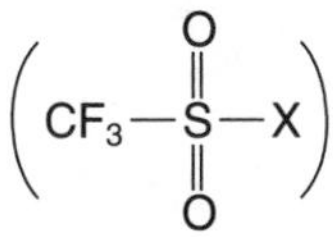

These groups are much better leaving groups than the hydroxy group because they are resonance stabilized. Alcohol molecules that are going to be reacted by S_N1 or S_N2 mechanisms are often first converted to their sulfonate esters to improve both the rate and yield of the reactions.

Formation of aldehydes and ketones. The oxidation of alcohols can lead to the formation of aldehydes and ketones. Aldehydes are formed from primary alcohols, while ketones are formed from secondary alcohols.

Because you can easily further oxidize aldehydes to carboxylic acids, you can only employ mild oxidizing agents and conditions in the formation of aldehydes. Typical mild oxidizing agents include manganese dioxide (MnO_2), Sarett-Collins reagent (CrO_3—$(C_5H_5N)_2$), and pyridinium chlorochromate (PCC),

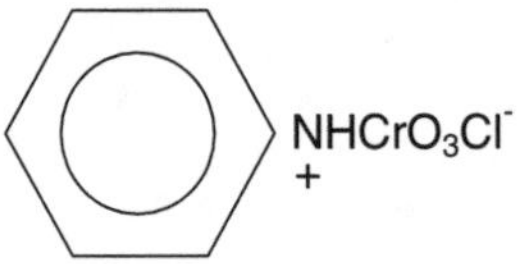

Following are several examples of the oxidation of primary alcohols:

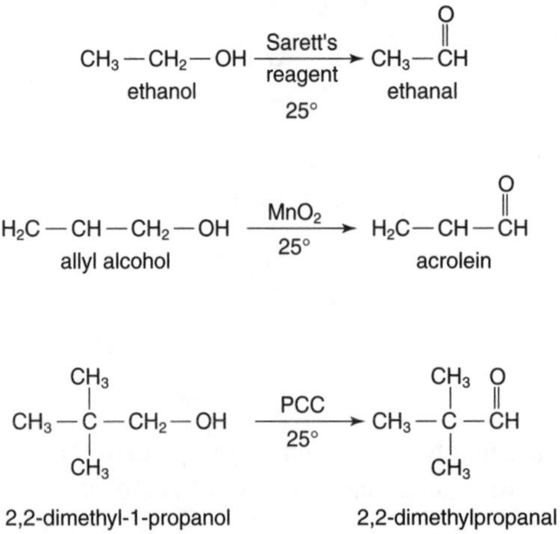

Because ketones are more resistant to further oxidation than aldehydes, you may employ stronger oxidizing agents and higher temperatures. Secondary alcohols are normally converted to ketones by reaction with potassium dichromate ($K_2Cr_2O_7$), potassium permanganate ($KMnO_4$), or chromium trioxide in acetic acid (CrO_3/CH_3COOH). Following are several examples of the oxidation of secondary alcohols:

$$CH_3-\overset{\overset{\displaystyle OH}{|}}{CH}-CH_3 \xrightarrow[\substack{H_2SO_4 \\ 70°}]{K_2Cr_2O_7} CH_3-\overset{\overset{\displaystyle O}{||}}{C}-CH$$

2-propanol   propanone

$$CH_3-CH_2-\overset{\overset{\displaystyle OH}{|}}{CH}-CH_3 \xrightarrow[\substack{H_2SO_4 \\ 80°}]{KMnO_4} CH_3-CH_2-\overset{\overset{\displaystyle O}{||}}{C}-CH_3$$

2-butanol   2-butanone

$$CH_3-CH_2-\overset{\overset{\displaystyle OH}{|}}{CH}-CH_3 \xrightarrow[\substack{CH_3COOH \\ 80°}]{CrO_3} CH_3-CH_2-\overset{\overset{\displaystyle O}{||}}{C}-CH$$

2-butanol   2-butanone

Carboxylic acid formation. Upon oxidation with strong oxidizing agents and high temperatures, primary alcohols completely oxidize to form carboxylic acids. The common oxidizing agents used for these conversions are concentrated potassium permanganate or concentrated potassium dichromate. Following are several examples of this type of oxidation:

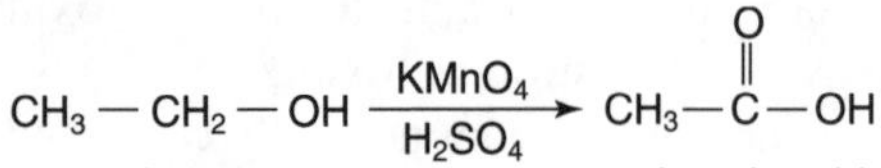

$$CH_3-CH_2-OH \xrightarrow[\substack{H_2SO_4 \\ 100°}]{KMnO_4} \overset{\overset{O}{\parallel}}{CH_3-C-OH}$$

ethanol   ethanoic acid

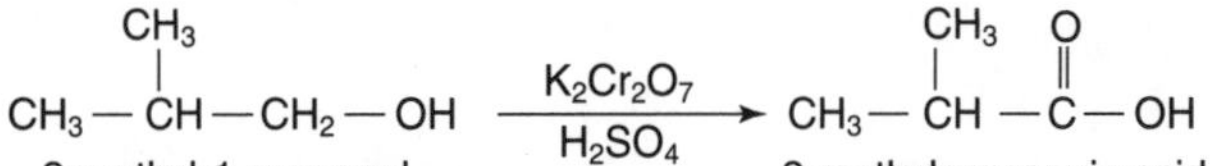

$$\overset{\overset{CH_3}{|}}{CH_3-CH-CH_2-OH} \xrightarrow[\substack{H_2SO_4 \\ 100°}]{K_2Cr_2O_7} \overset{\overset{CH_3\ \ O}{|\ \ \ \parallel}}{CH_3-CH-C-OH}$$

2-methyl-1-propanol   2-methylpropanoic acid

Ethers

Ethers are alkoxy (RO—)-substituted alkanes, alkenes, and alkynes. As with alcohols, only saturated carbon atoms may be substituted in alkenes and alkynes.

$$CH_3CH_2CH_2-\overset{..}{\underset{..}{O}}-CH_3 \qquad CH_2=CH-CH_2-\overset{..}{\underset{..}{O}}-CH_3$$

methyl propyl ether   allyl methyl ether

Nomenclature

Ethers are commonly named by listing the names of the groups attached to the oxygen atom and adding the word ether. Examples include:

$$CH_3-CH_2-\overset{..}{\underset{..}{O}}-CH_3 \qquad CH_3-\overset{..}{\underset{..}{O}}-CH=CH_2$$

ethyl methyl ether   methyl vinyl ether   methyl phenyl ether

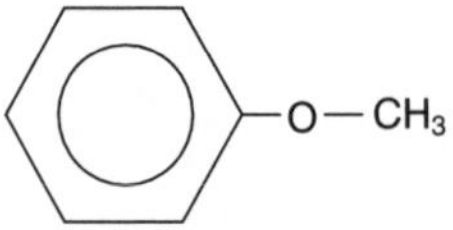

IUPAC nomenclature names ethers as alkoxy alkanes, alkoxy alkenes, or alkoxy alkynes. The group in the chain that has the greatest number of carbon atoms is designated the parent compound. In the case of aromatic ethers, the benzene ring is the parent compound.

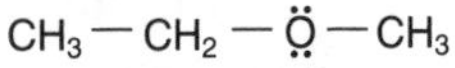

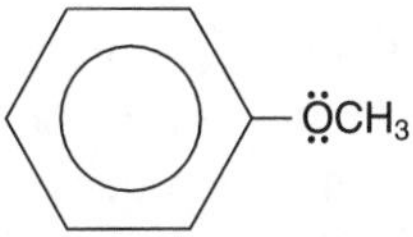

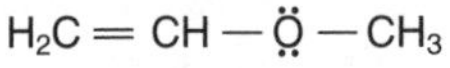

$CH_3-CH_2-\overset{\cdot\cdot}{\underset{\cdot\cdot}{O}}-CH_3$
methoxyethane

methoxybenzene

$H_2C=CH-\overset{\cdot\cdot}{\underset{\cdot\cdot}{O}}-CH_3$
methoxyethene

Cyclic ethers, oxygen-containing ring systems, are normally called by their common names.

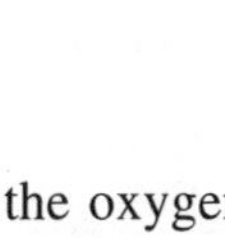

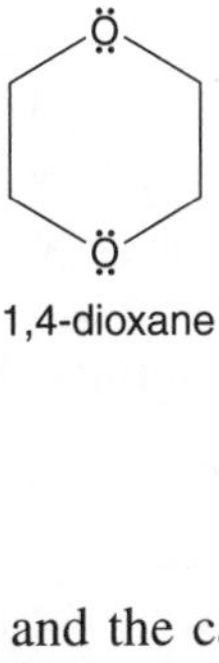

tetrahydrofuran   1,4-dioxane

Physical properties
The bonds between the oxygen atom and the carbon atoms of the alkyl groups in an ether molecule are polarized due to a difference in electronegativities between carbon and oxygen. In addition, the bond angle between the alkyl groups on the oxygen is 110°. These facts show that ether molecules must be dipoles (molecules having both a center of positive and negative charge) with weak polarities. Thus, the structure of ether is similar to that of water.

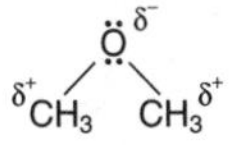

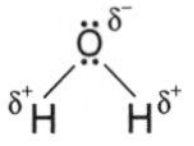

dimethyl ether   water

However, in water the hydrogen atoms have a greater partial positive charge than the hydrogen atoms on ether. In water, the charge is localized (only on) the hydrogens and not delocalized (spread throughout) as with the alkyl groups, so the charge is stronger in water than in ethers.

Like water, ether is capable of forming hydrogen bonds. However, because of the delocalized nature of the positive charge on the ether molecule's hydrogen atoms, the hydrogens cannot partake in hydrogen bonding. Thus, ethers only form hydrogen bonds to other molecules that have hydrogen atoms with strong partial positive charges. Therefore, ether molecules cannot form hydrogen bonds with other ether molecules. This leads to the high volatility of ethers. Ethers are capable, however, of forming hydrogen bonds to water, which accounts for the good solubility of low molecular weight ethers in water.

Table 5-1 shows boiling points for some simple ethers and the boiling points of alcohols of the same number of carbon atoms. Notice that due to the hydrogen bonding between alcohol molecules, all alcohols have appreciably higher boiling points than their isomeric ethers.

Table 5-1: Boiling Points of Simple Ethers and Alcohols

Ether	Boiling Point (°C)	Alcohol	Boiling Point (°C)
CH_3OCH_3 Dimethylether	-25	CH_3CH_2OH Ethanol	79
$CH_3OCH_2CH_3$ Methylethyether	11	$CH_3CH_2CH_2OH$ Propanol	82
$CH_3CH_2OCH_2CH_3$ Diethylether	35	$CH_3CH_2CH_2CH_2OH$ Butanol	117

Synthesis of Ethers

The sulfuric acid process and the Williamson method are both used to form ethers.

Sulfuric acid process

This method is used to make sterically hindered symmetrical ethers.

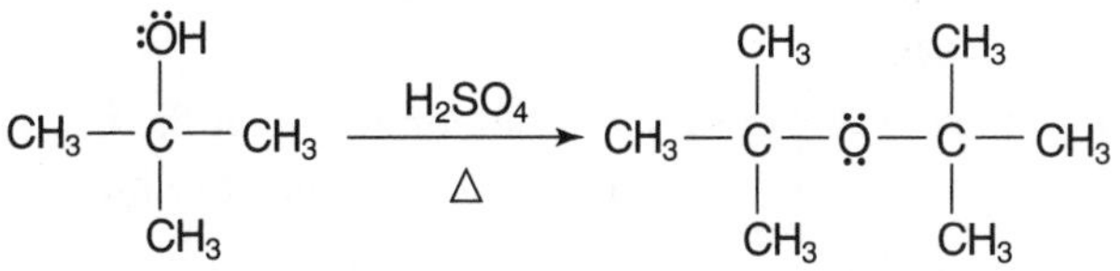

$$CH_3-\underset{\underset{CH_3}{|}}{\overset{\overset{:\ddot{O}H}{|}}{C}}-CH_3 \xrightarrow[\Delta]{H_2SO_4} CH_3-\underset{\underset{CH_3}{|}}{\overset{\overset{CH_3}{|}}{C}}-\ddot{O}-\underset{\underset{CH_3}{|}}{\overset{\overset{CH_3}{|}}{C}}-CH_3$$

The mechanism of the sulfuric acid process involves the following five steps.

1. Sulfuric acid dissociates, giving a proton plus the bisulfate ion.

$$H_2SO_4 \rightleftharpoons H^+ + HSO_4^-$$

2. The alcohol's oxygen atom is protonated via an acid-base reaction, leading to the formation of an oxonium ion.

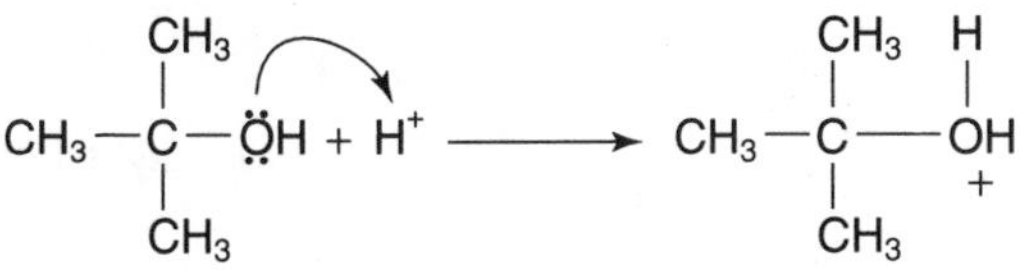

$$CH_3-\underset{\underset{CH_3}{|}}{\overset{\overset{CH_3}{|}}{C}}-\ddot{O}H + H^+ \longrightarrow CH_3-\underset{\underset{CH_3}{|}}{\overset{\overset{CH_3}{|}}{C}}\overset{\overset{H}{|}}{\underset{+}{-}}OH$$

an oxonium ion

3. The oxonium ion decomposes, generating a 3° carbocation and water. Because carbocations are planar, this decomposition destroys the steric hindrance effect that the t-butyl group created.

$$CH_3-\underset{\underset{CH_3}{|}}{\overset{\overset{CH_3}{|}}{C}}-\underset{+}{\overset{H}{\overset{|}{O}}}-H \longrightarrow CH_3-\underset{\underset{CH_3}{|}}{\overset{\overset{CH_3}{|}}{C^+}} + H\overset{..}{\underset{..}{O}}H$$

4. In this step, the acid-base reaction between the carbocation and a second molecule of alcohol takes place, which forms an oxonium ion.

$$CH_3-\underset{\underset{CH_3}{|}}{\overset{\overset{CH_3}{|}}{C^+}} + H\overset{..}{\underset{..}{O}}-\underset{\underset{CH_3}{|}}{\overset{\overset{CH_3}{|}}{C}}-CH_3 \longrightarrow CH_3-\underset{\underset{CH_3}{|}}{\overset{\overset{CH_3}{|}}{C}}-\underset{H}{\overset{+}{\overset{|}{O}}}-\underset{\underset{CH_3}{|}}{\overset{\overset{CH_3}{|}}{C}}-CH_3$$

5. The oxonium ion liberates a proton to yield the ether.

$$CH_3-\underset{\underset{CH_3}{|}}{\overset{\overset{CH_3}{|}}{C}}-\underset{H}{\overset{+}{\overset{|}{O}}}-\underset{\underset{CH_3}{|}}{\overset{\overset{CH_3}{|}}{C}}-CH_3 \longrightarrow CH_3-\underset{\underset{CH_3}{|}}{\overset{\overset{CH_3}{|}}{C}}-\overset{..}{\underset{..}{O}}-\underset{\underset{CH_3}{|}}{\overset{\overset{CH_3}{|}}{C}}-CH_3 + H^+$$

di-butyl ether

Williamson method

The Williamson ether synthesis proceeds via an S_N2 mechanism, in which an alkoxide ion displaces a halogen ion.

$$CH_3O^-Na^+ + CH_3Cl \longrightarrow CH_3OCH_3 + Na^+ + Cl^-$$

This method cannot be used with tertiary alkyl halides, because the competing elimination reaction predominates. The elimination reaction occurs because the rearward approach that is needed for an S_N2 mechanism is impossible due to steric hindrance. An S_N1 mechanism is likewise unfavored, because as the 3° carbon attempts to become a carbocation, the hydrogens on the adjacent carbons become acidic. Under these conditions, the alkoxide ion begins to show less nucleophilic character and, correspondingly, more basic character. This basic character leads to an acid-base reaction, which results in the generation of an elimination product (an alkene).

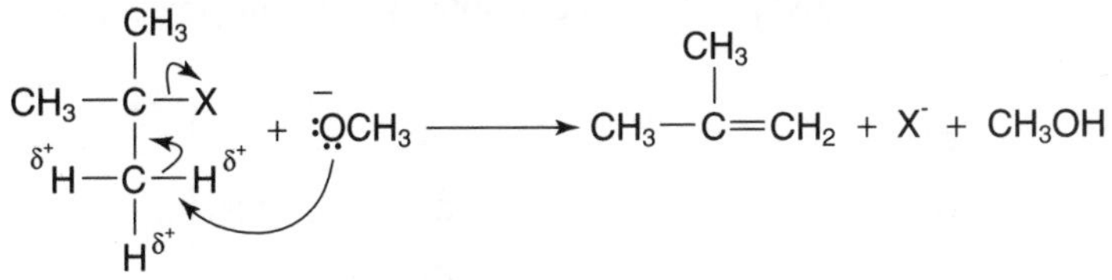

Reactions of Ethers

Although ethers are relatively inert toward reaction, they usually show good solvent properties for many nonpolar organic compounds. This strong dissolving power coupled with low reactivity makes ethers good solvents in which to run reactions.

An acid-catalyzed cleavage that occurs when hydriodic acid (HI) mixes with ethers is the most significant reaction that ethers experience. This reaction proceeds via a nucleophilic substitution mechanism. Primary and secondary alkyl ethers react by an S_N2 mechanism, while tertiary, benzylic, and alcylic ethers cleave by an S_N1 mechanism. A typical S_N2 reaction would be the reaction of ethylisopropyl ether with HI. The mechanism for this reaction is:

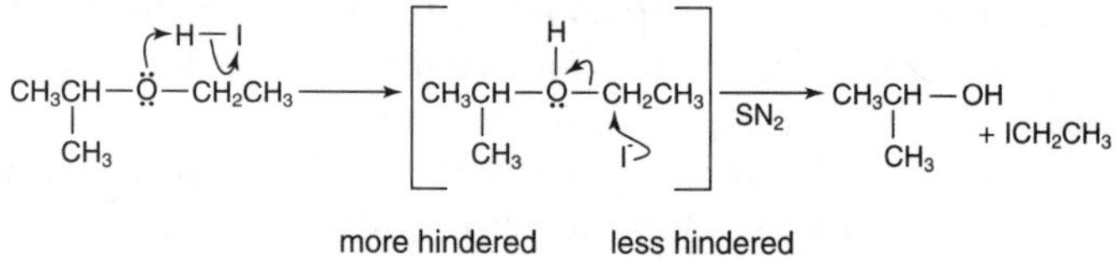

Notice that for S_N2 substitution, the alkyl halide came from the less sterically hindered group. For S_N1 type reactions, the alkyl halide forms from the fragment of the original molecule that forms the more stable cation. Thus, the reaction of t-butyl ethyl ether with HI gives t-butyl iodide and ethyl alcohol. The following mechanism occurs:

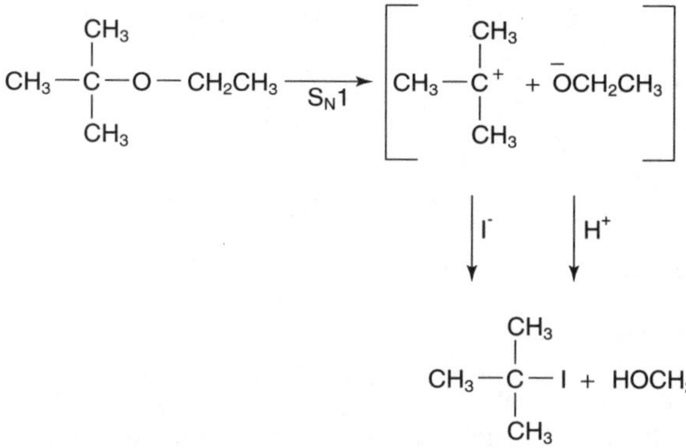

Notice that if the original ionization of the t-butyl ethyl ether formed a t-butoxide ion and an ethyl carbocation, this would be a less stable arrangement. (Remember, the order of stability of carbocations is 3° > 2° > 1°.)

Aldehydes

All aldehydes, except for formaldehyde, have a carbonyl group (C=O) attached to a hydrogen on one side and an alkyl or aryl group on the other side. Formaldehyde, the simplest aldehyde, has a hydrogen on both sides of the carbonyl group.

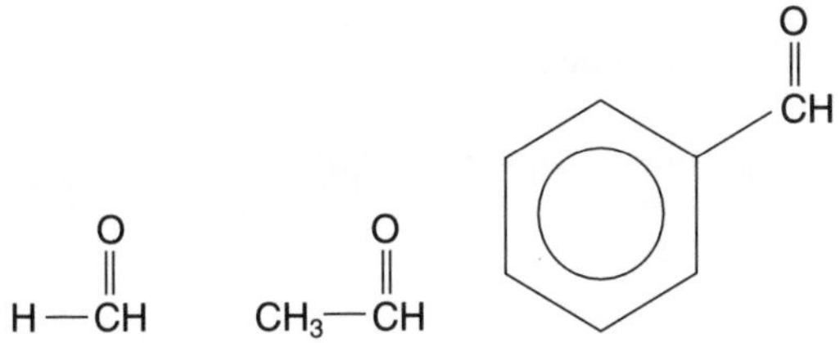

methanal   ethanal   benzaldehyde
(formaldehyde) (acetaldehyde)

Nomenclature

Aldehydes are named using either the common system or the IUPAC system. Aldehyde common names are derived from the common names of the corresponding carboxylic acid. For example:

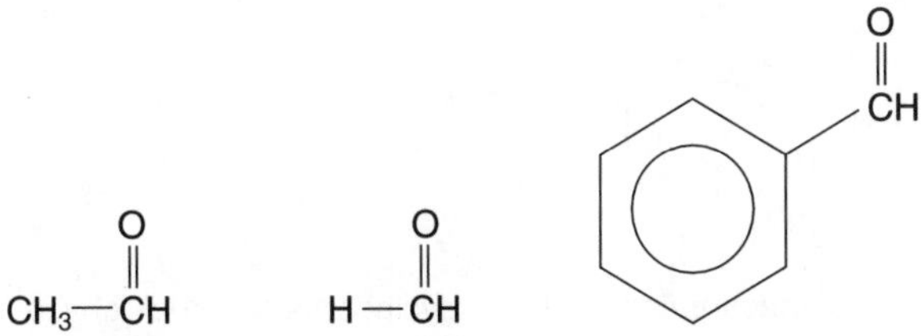

acetaldehyde   formaldehyde   benzaldehyde

The IUPAC system employs a series of rules to formulate compound names. The following rules explain how to name aliphatic aldehydes:

1. Pick out the longest continuous chain of carbon atoms that contains the carbonyl group.

2. The parent name comes from the alkane name of the same number of carbons.

3. Drop the -e ending of the alkane name and add -al.

4. Number the chain so that the carbonyl carbon has the lower possible number.

5. Locate and name substituents.

The following examples illustrate IUPAC naming:

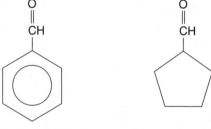

In the IUPAC system, cyclic, aliphatic, and aromatic aldehydes are named by adding the word carbaldehyde to the name of the ring system.

benzenecarbaldehyde   cyclopentanecarbaldehyde

Most chemists use the name benzaldehyde instead of benzenecarbaldehyde. Many texts and articles also use this alternate name.

Physical properties

Both aldehydes and ketones, which are described later in this chapter, have higher boiling points than hydrocarbons of similar chain lengths due to a greater degree of polarity and greater dipole-dipole interaction between molecules. Carbonyl groups cannot form strong intermolecular hydrogen bonds, so both aldehydes and ketones generally boil at lower temperatures than their alcohol analogues. However, aldehydes and ketones of low molecular weight do form strong hydrogen bonds with water, leading to good solubility.

Structure of the carbonyl group

In both aldehydes and ketones, the carbonyl group, with its attached hydrogen atom, alkyl group, or aryl group, lies in a plane with all bond angles close to 120°. The bond length of the carbon-oxygen double bond is significantly shorter than the average bond length of a carbon-oxygen single bond in alcohols or ethers. This factor makes the carbonyl group polar. Both aldehydes and ketones have dipole moments that are substantially greater than those of analogues with carbon-carbon double bonds. For example, 1-butene shows a dipole moment of 0.3 debye units (0.3 D), while propanol has a dipole moment of 2.5 D.

Synthesis of Aldehydes

Aldehydes can be prepared via a number of pathways. Some of the common methods are explained here.

The oxidation of primary alcohols

Primary alcohols can be oxidized by mild oxidizing agents, such as pyridinium chlorochromate (PCC), to yield aldehydes.

$$CH_3—CH_2—OH \xrightarrow[CH_2Cl_2]{C_6H_5\overset{+}{N}HCrO_3\overset{-}{Cl}PCC)} CH_3—\overset{\displaystyle O}{\overset{\|}{C}H}$$

ethanol   ethanal

The use of strong oxidizing agents would lead to the fully-oxidized product (carboxylic acid).

The reduction of acyl chlorides, esters, and nitriles

Reduction by mild reducing agents converts acyl chlorides, esters, and nitrites into aldehydes. The reducing agents of choice are usually lithium tri-tert-butoxy aluminum hydride (LATB—H) and diisobutylaluminum hydride (DIBAL—H). Following are the structures for these compounds:

$$Li^+ \left[H—\underset{\underset{\displaystyle OC(CH_3)_3}{|}}{\overset{\overset{\displaystyle O—C—(CH_3)_3}{|}}{Al}}—OC(CH_3)_3 \right]^-$$

$$H—\underset{\underset{\displaystyle CH_3}{|}}{\overset{\overset{\displaystyle CH_3}{|}}{Al}} \begin{matrix} CH_2CH—CH_3 \\ \\ CH_2CH—CH_3 \end{matrix}$$

lithium tri-tert-butoxyaluminum hydride
(LATB-H)

diisobutylaluminum hydride
(DIBAL-H)

Acyl chloride reduction

Acyl chlorides can be reduced by reacting them with lithium tri-tert-butoxyaluminum hydride at -78°C.

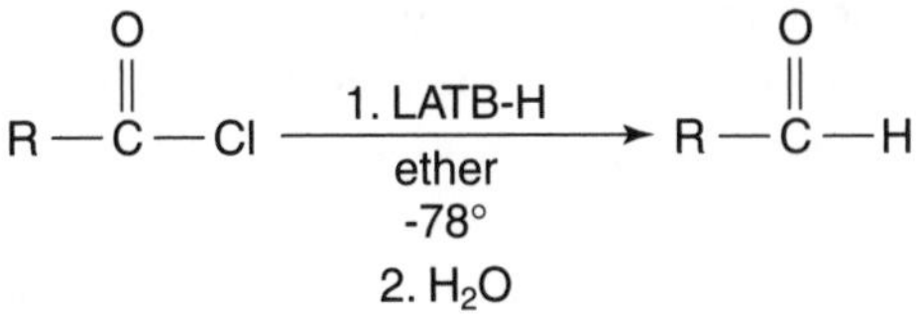

$$R-\overset{\overset{\displaystyle O}{\|}}{C}-Cl \xrightarrow[\substack{\text{ether} \\ -78° \\ \text{2. } H_2O}]{\text{1. LATB-H}} R-\overset{\overset{\displaystyle O}{\|}}{C}-H$$

Following is a typical reduction of an aromatic compound employing this reagent.

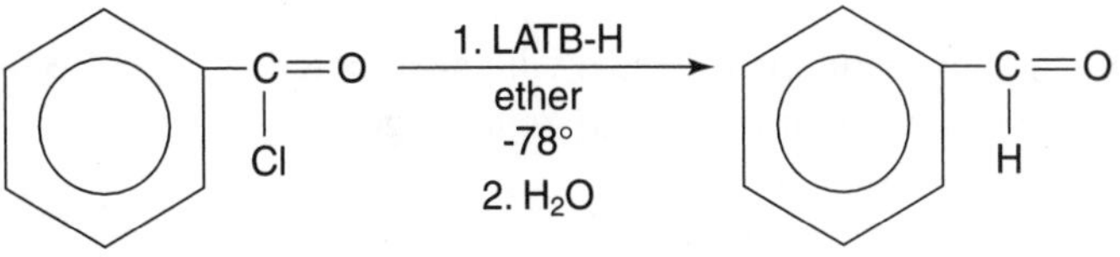

benzoyl chloride   benzaldehyde

The mechanism for acid chloride reduction proceeds via a hydride-ion transfer from the reducing agent to the acid chloride. The following steps summarize this mechanism.

1. An acid-base reaction occurs between an electron pair on the oxygen of the carbonyl group and the aluminum atom of the LATB—H.

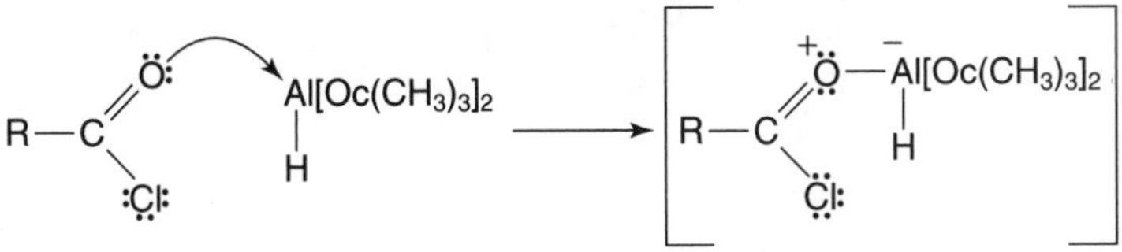

2. A hydride ion is transferred to the carbonyl carbon with a corresponding movement of the π electrons of the carbonyl group to the carbonyl oxygen.

$$
\left[\begin{array}{c} R-C \overset{\displaystyle \overset{+}{O}-Al[Oc(CH_3)_3]_2}{\underset{Cl}{\big|}} \underset{H}{} \end{array} \right] \longrightarrow \left[\begin{array}{c} :\overset{..}{O}-Al[Oc(CH_3)_3]_2 \\ R-\overset{|}{\underset{|}{C}}-H \\ Cl \end{array} \right]
$$

3. A chloride ion is liberated with anchiomeric assistance from an electron pair on the carbonyl oxygen.

$$
\left[\begin{array}{c} :\overset{..}{O}-Al[Oc(CH_3)_3]_2 \\ R-\overset{|}{\underset{|}{C}}-H \\ :\overset{..}{\underset{..}{Cl}}: \end{array} \right] \longrightarrow \left[\begin{array}{c} \overset{+..}{O}-Al[Oc(CH_3)_3]_2 \\ R-\overset{\|}{C}-H \end{array} \right] + :\overset{..}{\underset{..}{Cl}}:^-
$$

4. The aluminum complex is hydrolyzed by the addition of water to form the aldehyde.

$$
\left[\begin{array}{c} \overset{+..}{O}-Al[Oc(CH_3)_3]_2 \\ R-\overset{\|}{C}-H \end{array} \right] \xrightarrow{H_2O} R-\overset{\displaystyle O}{\overset{\|}{C}}-H
$$

Ester and nitrile reduction

You can use diisobutylaluminum hydride to reduce both esters and nitriles to aldehydes. Typical examples are the reduction of ethyl ethanoate (ethyl acetate) and ethanenitrile (acetonitrile) to ethanal (acetaldehyde).

$$CH_3 - \overset{\overset{\displaystyle \ddot{O}:}{\|}}{C} - \ddot{\underset{\cdot\cdot}{O}} - CH_2 - CH_3 \xrightarrow[\text{Hexane}]{\text{1. DIBAL-H}} CH_3 - \overset{\overset{\displaystyle \ddot{O}:}{\|}}{CH}$$

2. H_2O

$$CH_3 - C \equiv N \xrightarrow[\text{Hexane}]{\text{1. DIBAL-H}} CH_3 - \overset{\overset{\displaystyle \ddot{O}:}{\|}}{CH}$$

2. H_2O

The mechanism for both of these reactions is very similar to the mechanism for the reduction of acyl chlorides by LATB—H. The first step is an acid-base reaction between an unshared electron pair on oxygen or nitrogen with the aluminum atom of the DIBAL—H. The second step is the transfer of a hydride ion from the DIBAL—H to the carbon atom of the carbonyl or nitrile group. The last step is the hydrolysis of the aluminum complex to form the aldehyde.

Ozonolysis of alkenes
Alkenes in which the carbon(s) of the double bond possess one or more hydrogen atoms react with ozone (O_3) to generate aldehydes. The reaction of propene with ozone to form acetaldehyde and formaldehyde illustrates this method of preparation.

$$CH_3 - CH = CH_2 \xrightarrow{O_3} CH_3 - \underset{\underset{\displaystyle O - O}{\diagdown \diagup}}{\overset{\overset{\displaystyle O}{\diagup \diagdown}}{CH}} CH_2 \xrightarrow{H^+} CH_3 - \overset{\overset{\displaystyle O}{\|}}{CH} + H_2C = O$$

an ozonide

Hydroboration of terminal alkynes

Terminal alkynes react rapidly with borane to produce an intermediate compound that is easily oxidized to an aldehyde. For example, you can produce pentenal by reacting pentyne with borane and oxidizing the resulting intermediate with aqueous hydrogen peroxide.

Ketones

Ketones are compounds in which an oxygen atom is bonded to a carbon atom, which is itself bonded to two or more carbon atoms.

$$CH_3—CH_2—\overset{\overset{\displaystyle O}{\|}}{C}—CH_3$$

2-butanone
(methyl ethyl ketone)

1-phenylethanone
(methyl phenyl ketone)

diphenyl methanone
(diphenyl ketone)

Nomenclature

Like aldehydes, ketones can be named using either the common system or the IUPAC system. In the common system, ketones names are created by naming the groups attached to the carbonyl carbon and then adding the word *ketone*. Following are several examples:

$$CH_3—\overset{\overset{\displaystyle CH_3}{|}\atop\overset{\displaystyle CH_2}{|}}{C}=O$$

ethyl methyl ketone

$$CH_3—CH_2—CH_2—C=O$$

phenyl propyl ketone

phenyl p-tolyl ketone

The five IUPAC rules previously outlined in the "Nomenclature" section for aldehydes also apply to ketones, with one exception: After dropping the -e ending of the alkane name, you add -one for ketones (rather than -al, which designates aldehydes).

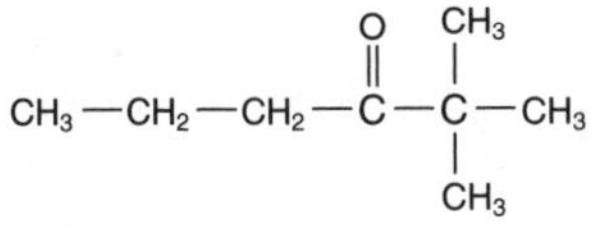

2,2-dimethyl-3-hexanone 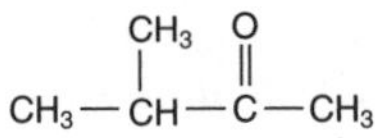   3-methyl-2-butanone

In the IUPAC system, aromatic ketones are considered benzene-substituted aliphatic ketones.

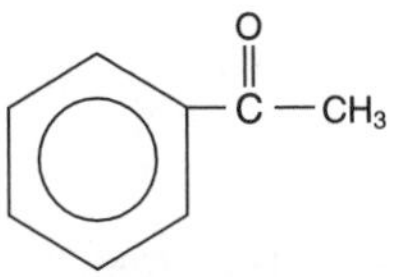

1-phenylethanone 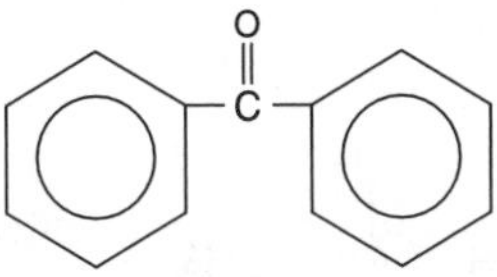   diphenyl methanone

Many aromatic compounds retain their common names in the IUPAC system. Following are two such examples:

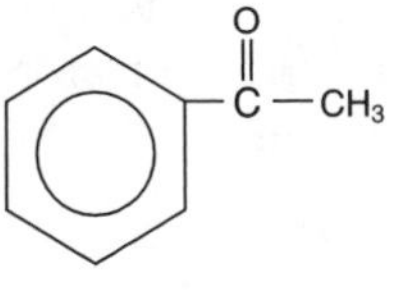

acetophenone 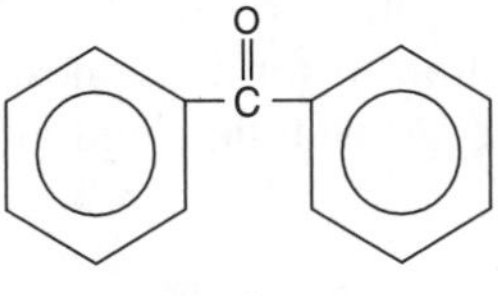   benzophenone

Synthesis of Ketones

Like aldehydes, ketones can be prepared in a number of ways. The following sections detail some of the more common preparation methods: the oxidation of secondary alcohols, the hydration of alkynes, the ozonolysis of alkenes, Friedel-Crafts acylation, the use of lithium dialkylcuprates, and the use of a Grignard reagent.

Oxidation of secondary alcohols

The oxidation of secondary alcohols to ketones may be carried out using strong oxidizing agents, because further oxidation of a ketone occurs with great difficulty. Normal oxidizing agents include potassium dichromate ($K_2Cr_2O_7$) and chromic acid (H_2CrO_4). The conversion of 2-propanol to 2-propanone illustrates the oxidation of a secondary alcohol.

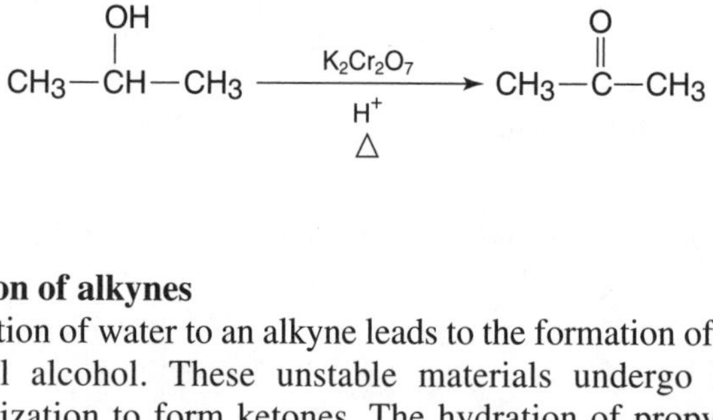

$$CH_3-\underset{\underset{OH}{|}}{CH}-CH_3 \xrightarrow[\underset{\Delta}{H^+}]{K_2Cr_2O_7} CH_3-\underset{\overset{O}{||}}{C}-CH_3$$

Hydration of alkynes

The addition of water to an alkyne leads to the formation of an unstable vinyl alcohol. These unstable materials undergo keto-enol tautomerization to form ketones. The hydration of propyne forms 2-propanone, as the following figure illustrates.

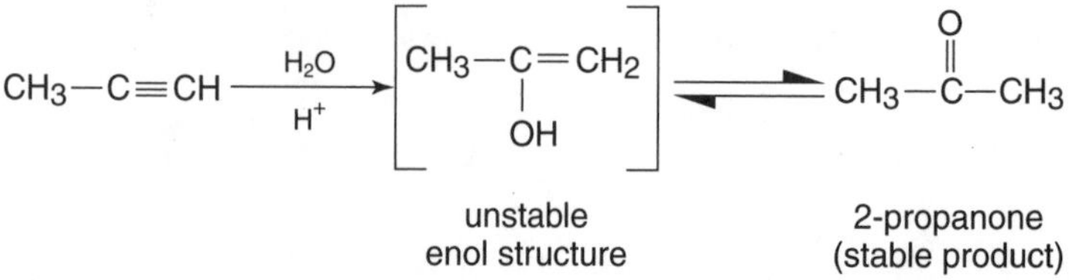

$$CH_3-C\equiv CH \xrightarrow[H^+]{H_2O} \left[CH_3-\underset{\underset{OH}{|}}{C}=CH_2 \right] \rightleftharpoons CH_3-\underset{\overset{O}{||}}{C}-CH_3$$

unstable
enol structure

2-propanone
(stable product)

Ozonolysis of alkenes

When one or both alkene carbons contain two alkyl groups, ozonolysis generates one or two ketones. The ozonolysis of 1,2-dimethyl propene produces both 2-propanone (a ketone) and ethanal (an aldehyde).

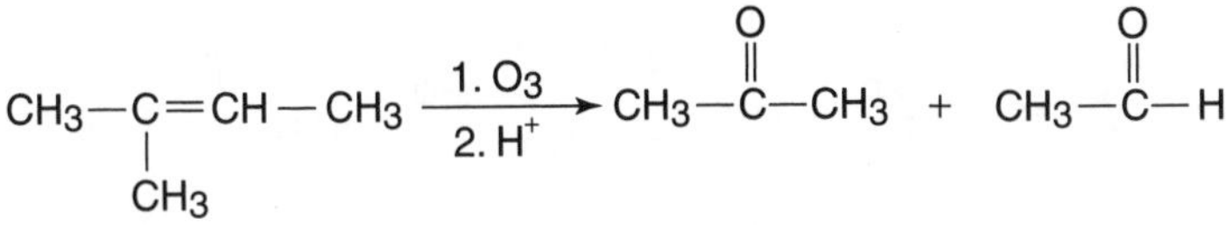

$$CH_3-\underset{\underset{CH_3}{|}}{C}=CH-CH_3 \xrightarrow[\text{2. } H^+]{\text{1. } O_3} CH_3-\overset{\overset{O}{||}}{C}-CH_3 \ + \ CH_3-\overset{\overset{O}{||}}{C}-H$$

Friedel-Crafts acylation

Friedel-Crafts acylations (see Chapter 2) are used to prepare aromatic ketones. The preparation of acetophenone from benzene and acetyl chloride is a typical Friedel-Crafts acylation.

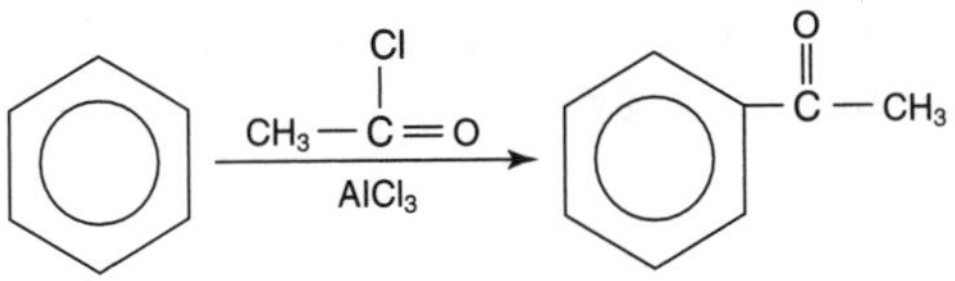

Lithium dialkylcuprates

The addition of a lithium dialkylcuprate (Gilman reagent) to an acyl chloride at low temperatures produces a ketone. This method produces a good yield of acetophenone.

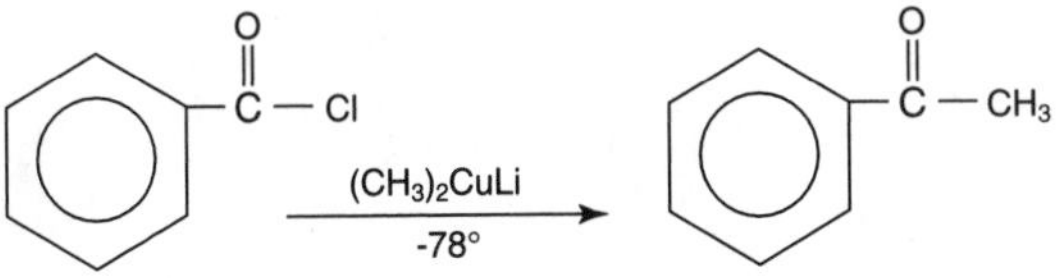

Grignard reagents

Hydrolysis of the salt formed by reacting a Grignard reagent with a
nitrile produces good ketone yields. For example, you can prepare
acetone by reacting the Grignard reagent methyl magnesium bromide
(CH_3MgBr) with methyl nitrile $(CH_3C\equiv N)$.

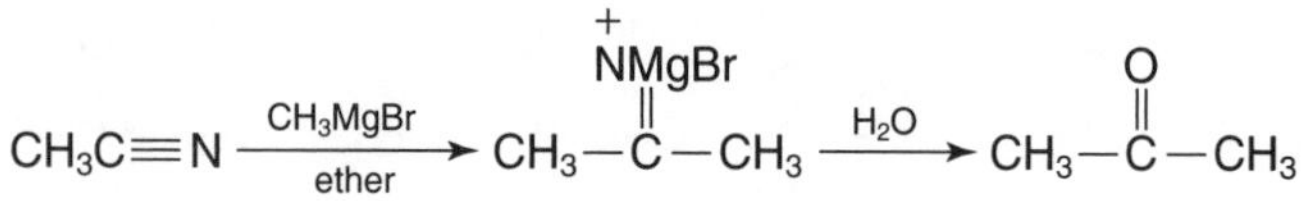

$$CH_3C\equiv N \xrightarrow[\text{ether}]{CH_3MgBr} CH_3-\overset{\overset{+}{NMgBr}}{\underset{}{\overset{\|}{C}}}-CH_3 \xrightarrow{H_2O} CH_3-\overset{O}{\overset{\|}{C}}-CH_3$$

Reactions of Aldehydes and Ketones

Aldehydes and ketones undergo a variety of reactions that lead to many
different products. The most common reactions are nucleophilic addition reactions, which lead to the formation of alcohols, alkenes, diols,
cyanohydrins $(RCH(OH)C\equiv N)$, and imines $(R_2C=NR)$, to mention
a few representative examples.

Reactions of carbonyl groups

The main reactions of the carbonyl group are nucleophilic additions
to the carbon-oxygen double bond. As shown below, this addition
consists of adding a nucleophile and a hydrogen across the carbon-
oxygen double bond.

$$CH_3 \quad H + R \quad R \longrightarrow R \quad R \longrightarrow R-\overset{:\ddot{O}:^-}{\underset{CH_3-\ddot{O}:}{\overset{|}{C}}}-R + H^+ \longrightarrow R-\overset{:\ddot{O}H}{\underset{CH_3-\ddot{O}:}{\overset{|}{C}}}-R$$

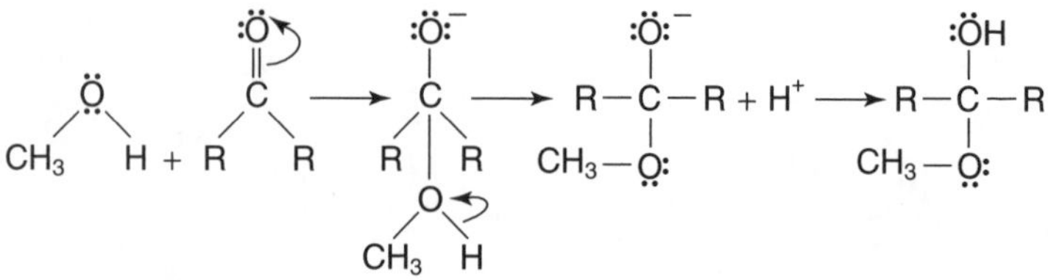

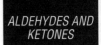

Due to differences in electronegativities, the carbonyl group is polarized. The carbon atom has a partial positive charge, and the oxygen atom has a partially negative charge.

$$\underset{R \overset{\delta^+}{} R}{\overset{\overset{\displaystyle O^{\delta^-}}{\|}}{C}}$$

Aldehydes are usually more reactive toward nucleophilic substitutions than ketones because of both steric and electronic effects. In aldehydes, the relatively small hydrogen atom is attached to one side of the carbonyl group, while a larger R group is affixed to the other side. In ketones, however, R groups are attached to both sides of the carbonyl group. Thus, steric hindrance is less in aldehydes than in ketones.

Electronically, aldehydes have only one R group to supply electrons toward the partially positive carbonyl carbon, while ketones have two electron-supplying groups attached to the carbonyl carbon. The greater amount of electrons being supplied to the carbonyl carbon, the less the partial positive charge on this atom and the weaker it will become as a nucleus.

Addition of water
The addition of water to an aldehyde results in the formation of a hydrate.

$$CH_3 - \overset{\overset{\displaystyle O}{\|}}{C} - H \xrightarrow[\text{H}^+]{\text{H}_2\text{O}} CH_3 - \underset{\underset{\displaystyle OH}{|}}{\overset{\overset{\displaystyle OH}{|}}{C}} - H$$

hydrate

The formation of a hydrate proceeds via a nucleophilic addition mechanism.

1. Water, acting as a nucleophile, is attracted to the partially positive carbon of the carbonyl group, generating an oxonium ion.

$$CH_3-\overset{\overset{\displaystyle\delta^-}{\ddot{\overset{..}{O}}:}}{\underset{\delta^+}{C}}-H + H\overset{\overset{\displaystyle\ddot{\overset{..}{O}}}{}}{\diagdown}H \longrightarrow CH_3-\underset{\underset{H\quad H}{\overset{..}{\underset{..}{\overset{+}{O}}}}}{\overset{:\overset{..}{\overset{..}{O}}:^-}{C}}-H$$

2. The oxonium ion liberates a hydrogen ion that is picked up by the oxygen anion in an acid-base reaction.

$$CH_3-\underset{\underset{H\quad H}{\overset{..}{\underset{..}{\overset{+}{O}}}:}}{\overset{:\overset{..}{\overset{..}{O}}:^-}{C}}-H \rightleftharpoons CH_3-\underset{:\overset{..}{O}H}{\overset{:\overset{..}{O}H}{C}}-H$$

Small amounts of acids and bases catalyze this reaction. This occurs because the addition of acid causes a protonation of the oxygen of the carbonyl group, leading to the formation of a full positive charge on the carbonyl carbon, making the carbon a good nucleus. Adding hydroxyl ions changes the nucleophile from water (a weak nucleophile) to a hydroxide ion (a strong nucleophile). Ketones usually do not form stable hydrates.

Addition of alcohol

Reactions of aldehydes with alcohols produce either **hemiacetals** (a functional group consisting of one —OH group and one —OR group bonded to the same carbon) or **acetals** (a functional group consisting of two —OR groups bonded to the same carbon), depending upon conditions. Mixing the two reactants together produces the hemiacetal. Mixing the two reactants with hydrochloric acid produces an acetal. For example, the reaction of methanol with ethanal produces the following results:

$$CH_3-\overset{\overset{\text{O}}{\|}}{C}H + CH_3-OH \xrightarrow{\Delta} CH_3-\overset{\overset{\text{OH}}{|}}{\underset{\underset{\text{OCH}_3}{|}}{C}}-H$$

a hemiacetal

$$CH_3-\overset{\overset{\text{O}}{\|}}{C}H + CH_3-OH + HCl \xrightarrow{\Delta} CH_3-\overset{\overset{\text{OCH}_3}{|}}{\underset{\underset{\text{OCH}_3}{|}}{C}}-H$$

an acetal

A nucleophilic substitution of an OH group for the double bond of the carbonyl group forms the hemiacetal through the following mechanism:

1. An unshared electron pair on the alcohol's oxygen atom attacks the carbonyl group.

$$CH_3-\overset{\overset{\text{Ö:}}{\|}}{C}-H + H\overset{..}{\underset{..}{O}}-CH_3 \longrightarrow CH_3-\overset{\overset{:\overset{..}{O}^-}{|}}{\underset{\underset{\text{H}}{|}}{C}}-\overset{\overset{+}{\overset{..}{O}}}{\underset{\underset{\text{H}}{|}}{}}-CH_3$$

2. The loss of a hydrogen ion to the oxygen anion stabilizes the oxonium ion formed in Step 1.

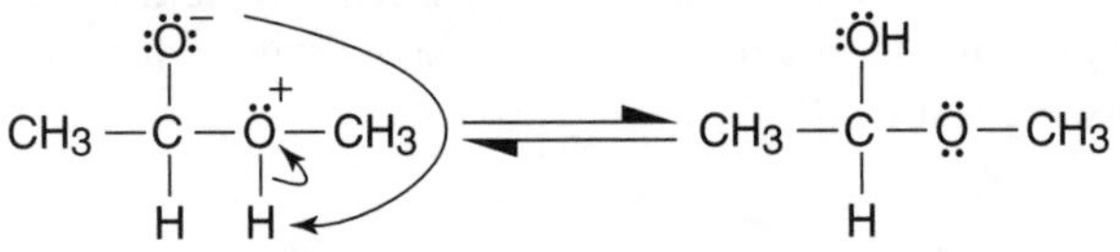

$$CH_3 - \underset{\underset{H}{|}}{\overset{\overset{:\ddot{O}:^-}{|}}{C}} - \overset{+}{\underset{\underset{H}{}}{\ddot{O}}} - CH_3 \;\rightleftharpoons\; CH_3 - \underset{\underset{H}{|}}{\overset{\overset{:\ddot{O}H}{|}}{C}} - \ddot{O} - CH_3$$

The addition of acid to the hemiacetal creates an acetal through the following mechanism:

1. The proton produced by the dissociation of hydrochloric acid protonates the alcohol molecule in an acid-base reaction.

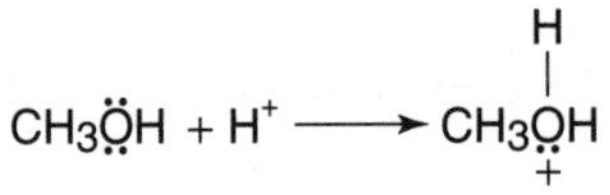

$$CH_3\ddot{O}H + H^+ \longrightarrow CH_3\underset{+}{\overset{\overset{H}{|}}{O}}H$$

2. An unshared electron pair from the hydroxyl oxygen of the hemiacetal removes a proton from the protonated alcohol.

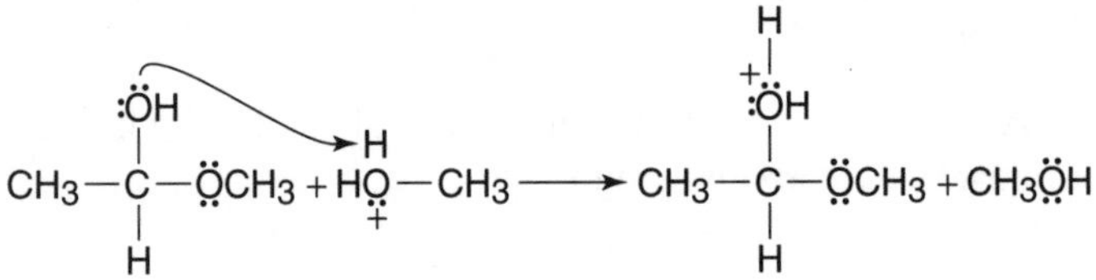

$$CH_3 - \underset{\underset{H}{|}}{\overset{\overset{:\ddot{O}H}{|}}{C}} - \ddot{O}CH_3 + H\underset{+}{\overset{\overset{H}{}}{\ddot{O}}} - CH_3 \longrightarrow CH_3 - \underset{\underset{H}{|}}{\overset{\overset{\overset{+}{:\ddot{O}}H}{|}}{C}} - \ddot{O}CH_3 + CH_3\ddot{O}H$$

3. The oxonium ion is lost from the hemiacetal as a molecule of water.

$$CH_3-\overset{\overset{+}{O}H}{\underset{H}{\overset{|}{C}}}-\overset{..}{O}CH_3 \longrightarrow CH_3-\overset{}{\underset{H}{\overset{|}{C}}}=\overset{+}{\overset{..}{O}}CH_3 + H_2\overset{..}{O}:$$

4. A second molecule of alcohol attacks the carbonyl carbon that is forming the protonated acetal.

$$CH_3-\overset{}{\underset{H}{\overset{|}{C}}}=\overset{+}{\overset{..}{O}}CH_3 + CH_3\overset{..}{O}H \longrightarrow CH_3-\overset{\overset{CH_3}{\underset{|}{\overset{+}{\overset{..}{O}H}}}}{\underset{H}{\overset{|}{C}}}-\overset{..}{O}CH_3$$

5. The oxonium ion loses a proton to an alcohol molecule, liberating the acetal.

$$CH_3-\overset{\overset{CH_3}{\underset{|}{\overset{+}{\overset{..}{O}H}}}}{\underset{H}{\overset{|}{C}}}-\overset{..}{O}CH_3 + CH_3\overset{..}{O}H \longrightarrow CH_3-\overset{\overset{:\overset{..}{O}CH_3}{|}}{\underset{H}{\overset{|}{C}}}-\overset{..}{O}CH_3 + CH_3\overset{\overset{H}{\underset{+}{\overset{|}{O}}}}{H}$$

Stability of acetals

Acetal formation reactions are reversible under acidic conditions but not under alkaline conditions. This characteristic makes an acetal an ideal protecting group for aldehyde molecules that must undergo further reactions. A **protecting group** is a group that is introduced into a molecule to prevent the reaction of a sensitive group while a reaction is carried out at some other site in the molecule. The protecting group must have the ability to easily react back to the original group from which it was formed. An example is the protection of an aldehyde group in a molecule so that an ester group can be reduced to an alcohol.

$$\underset{H}{\overset{O}{\parallel}}C-CH_2-\overset{O}{\overset{\parallel}{C}}-OCH_3 \xrightarrow[\Delta]{\overset{CH_3OH}{H^+}} H-\overset{OCH_3}{\underset{OCH_3}{\overset{|}{\underset{|}{C}}}}-CH_2-\overset{O}{\overset{\parallel}{C}}-O-CH_3$$

In the previous reaction, the aldehyde group is converted into an acetal group, thus preventing reaction at this site when further reactions are run on the rest of the molecule.

$$\xrightarrow[\substack{\text{ether} \\ 2.\ H_2O}]{1.\ LiAlH_4} H-\overset{OCH_3}{\underset{OCH_3}{\overset{|}{\underset{|}{C}}}}-CH_2-CH_2OH \xrightarrow[\Delta]{\overset{H^+}{H_2O}} \underset{H}{\overset{O}{\parallel}}C-CH_2CH_2OH$$

Notice in the previous reaction that the ketone carbonyl group has been reduced to an alcohol by reaction with LiAlH.. The protected aldehyde group has not been reduced. Hydrolysis of the reduction product recreates the original aldehyde group in the final product.

Addition of hydrogen cyanide
The addition of hydrogen cyanide to a carbonyl group of an aldehyde or most ketones produces a cyanohydrin. Sterically hindered ketones, however, don't undergo this reaction.

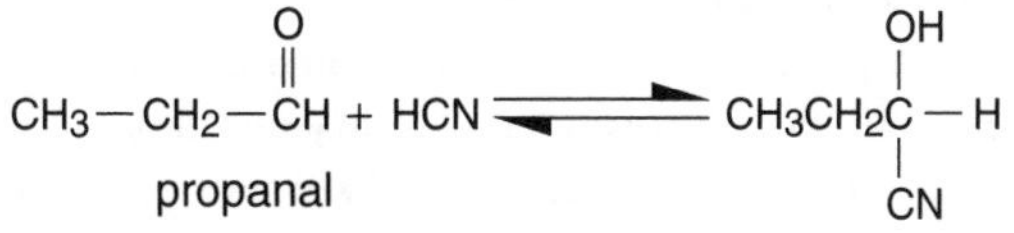

$$\underset{\text{propanal}}{CH_3-CH_2-\overset{\overset{\displaystyle O}{\|}}{CH}} + HCN \; \underset{\longleftarrow}{\overrightarrow{\hspace{2cm}}} \; CH_3CH_2\underset{\underset{\displaystyle CN}{|}}{\overset{\overset{\displaystyle OH}{|}}{C}}-H$$

$$\underset{\text{acetone}}{CH_3-\overset{\overset{\displaystyle O}{\|}}{C}-CH_3} + HCN \; \underset{\longleftarrow}{\overrightarrow{\hspace{2cm}}} \; CH_5\underset{\underset{\displaystyle CN}{|}}{\overset{\overset{\displaystyle OH}{|}}{C}}-CH_3$$

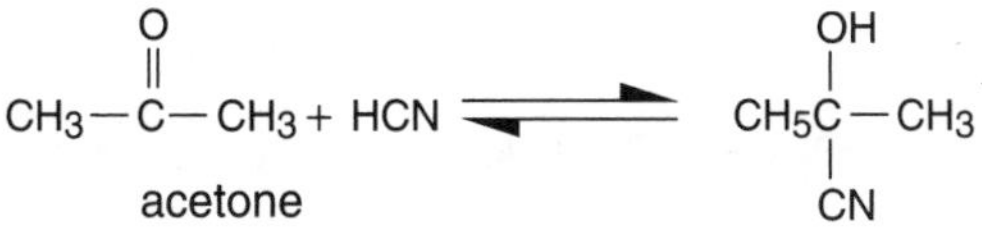

The mechanism for the addition of hydrogen cyanide is a straightforward nucleophilic addition across the carbonyl carbony oxygen bond.

Addition of ylides (the Wittig reaction)

The reaction of aldehydes or ketones with phosphorus ylides produces alkenes of unambiguous double-bond locations. Phosphorous ylides are prepared by reacting a phosphine with an alkyl halide, followed by treatment with a base. Ylides have positive and negative charges on adjacent atoms.

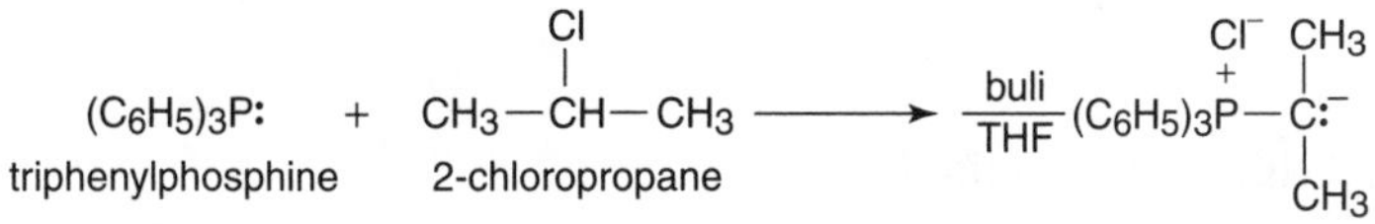

$$\underset{\text{triphenylphosphine}}{(C_6H_5)_3P\!:} \quad + \quad \underset{\text{2-chloropropane}}{CH_3-\overset{\overset{\displaystyle Cl}{|}}{CH}-CH_3} \quad \longrightarrow \quad \overset{\text{buli}}{\underset{\text{THF}}{\longrightarrow}} (C_6H_5)_3\overset{+}{P}-\underset{\underset{\displaystyle CH_3}{|}}{\overset{\overset{\displaystyle Cl^- \; CH_3}{|}}{C}\!:^-}$$

The following illustration shows the preparation of 2-methylbutene by a Wittig reaction.

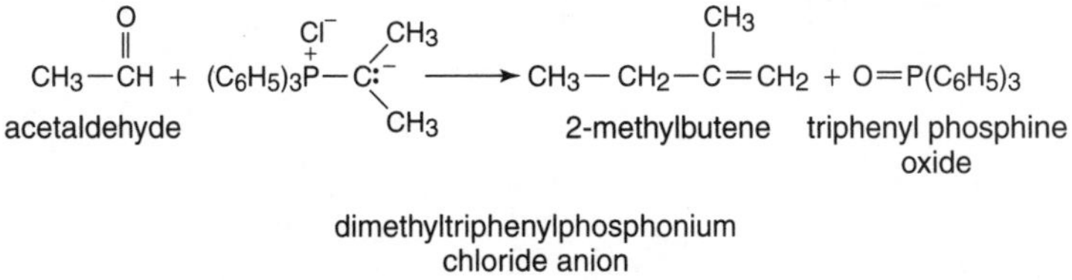

$$\underset{\text{acetaldehyde}}{CH_3-\overset{\overset{\displaystyle O}{\|}}{CH}} + \underset{\underset{\displaystyle CH_3}{}}{(C_6H_5)_3\overset{+}{P}-\overset{\overset{\displaystyle Cl^- \; CH_3}{}}{C\!:^-}} \longrightarrow \underset{\text{2-methylbutene}}{CH_3-CH_2-\overset{\overset{\displaystyle CH_3}{|}}{C}\!=CH_2} + \underset{\substack{\text{triphenyl phosphine} \\ \text{oxide}}}{O\!=\!P(C_6H_5)_3}$$

dimethyltriphenylphosphonium
chloride anion

Addition of organometallic reagents

Grignard reagents, organolithium compounds, and sodium alkynides react with formaldehyde to produce primary alcohols, all other aldehydes to produce secondary alcohols, and ketones to produce tertiary alcohols.

$$\overset{\delta^-}{R}-\overset{\delta^+}{MgX} + R-\overset{\overset{\textstyle O}{\|}}{C}-H(R) \xrightarrow{H^+} R-\overset{\overset{\textstyle OH}{|}}{\underset{\underset{\textstyle R}{|}}{C}}-H(R)$$

$$\overset{\delta^-}{R}-\overset{\delta^+}{Li} + R-\overset{\overset{\textstyle O}{\|}}{C}-H(R) \xrightarrow{H^+} R-\overset{\overset{\textstyle OH}{|}}{\underset{\underset{\textstyle H}{|}}{C}}-H(R)$$

$$R-C\equiv C^-Na^+ + R-\overset{\overset{\textstyle O}{\|}}{C}-H(R) \xrightarrow{H^+} R-C\equiv C-\overset{\overset{\textstyle OH}{|}}{\underset{\underset{\textstyle H}{|}}{C}}-H(R)$$

Addition of ammonia derivatives

Aldehydes and ketones react with primary amines to form a class of compounds called imines.

$$R-\overset{\overset{\textstyle O}{\|}}{C}-H(R) + H_2N-R' \rightleftharpoons R-N=C\overset{\textstyle H}{\underset{\textstyle H}{\big\langle}} + H_2O$$

The mechanism for imine formation proceeds through the following steps:

1. An unshared pair of electrons on the nitrogen of the amine is attracted to the partial-positive carbon of the carbonyl group.

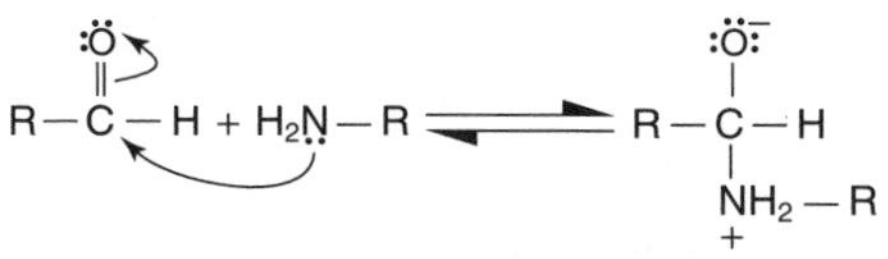

$$R-\overset{\overset{\ddot{O}}{\|}}{C}-H + H_2\ddot{N}-R \rightleftharpoons R-\overset{:\ddot{O}:^-}{\underset{\underset{+}{NH_2-R}}{C}}-H$$

2. A proton is transferred from the nitrogen to the oxygen anion.

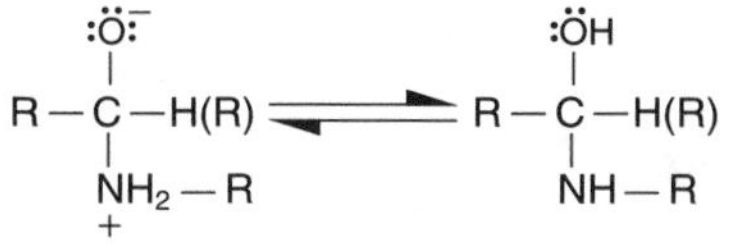

$$R-\overset{:\ddot{O}:^-}{\underset{\underset{+}{NH_2-R}}{C}}-H(R) \rightleftharpoons R-\overset{:\ddot{O}H}{\underset{NH-R}{C}}-H(R)$$

3. The hydroxy group is protonated to yield an oxonium ion, which easily liberates a water molecule.

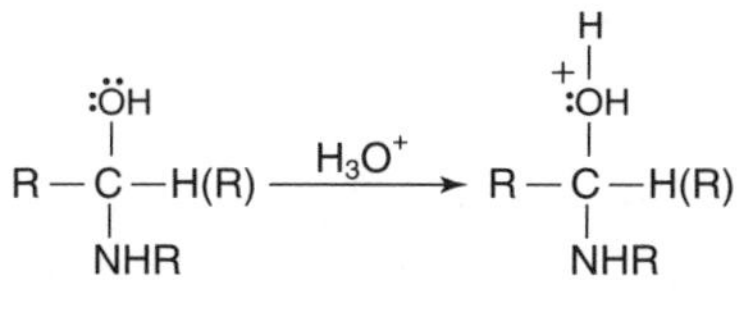

$$R-\overset{:\ddot{O}H}{\underset{NHR}{C}}-H(R) \xrightarrow{H_3O^+} R-\overset{\overset{+}{:}OH}{\underset{NHR}{C}}-H(R)$$

4. An unshared pair of electrons on the nitrogen migrate toward the positive oxygen, causing the loss of a water molecule.

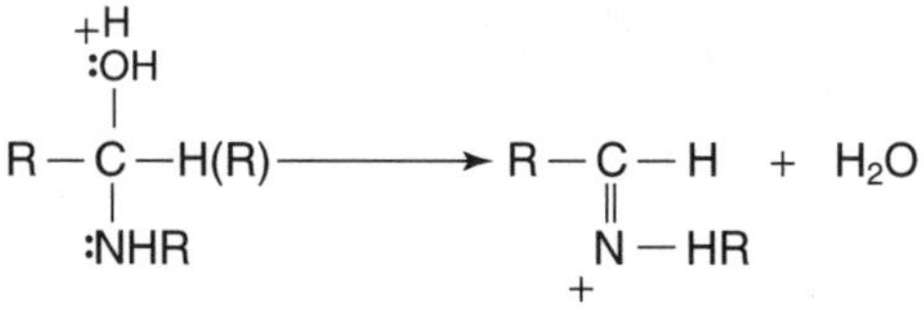

$$R-\overset{\overset{+H}{:OH}}{\underset{:NHR}{C}}-H(R) \longrightarrow R-\overset{\|}{\underset{\underset{+}{N-HR}}{C}}-H + H_2O$$

5. A proton from the positively charged nitrogen is transferred to water, leading to the imine's formation.

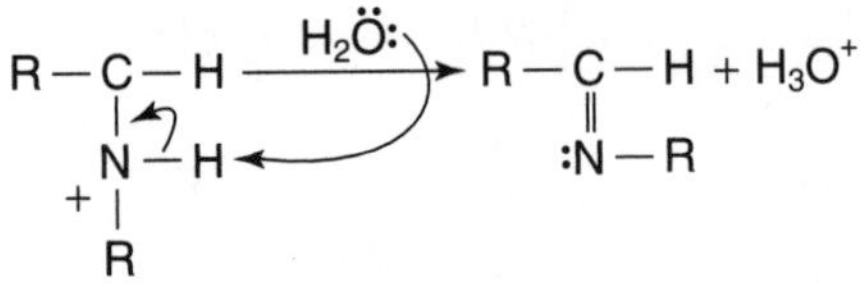

$$R-\overset{\displaystyle |}{\underset{\displaystyle \underset{R}{\overset{+}{N}-H}}{C}}-H \xrightarrow{\;H_2\ddot{O}:\;} R-\overset{\displaystyle \parallel}{\underset{\displaystyle :N-R}{C}}-H + H_3O^+$$

Imines of aldehydes are relatively stable while those of ketones are unstable. Derivatives of imines that form stable compounds with aldehydes and ketones include phenylhydrazine, 2,4-dinitrophenylhydrazine, hydroxylamine, and semicarbazide.

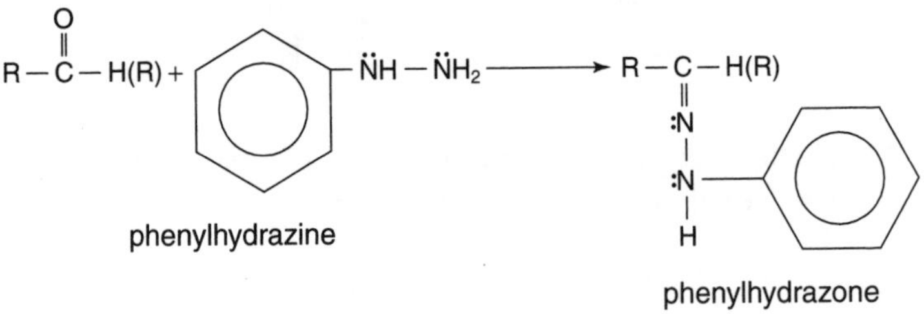

phenylhydrazine

phenylhydrazone

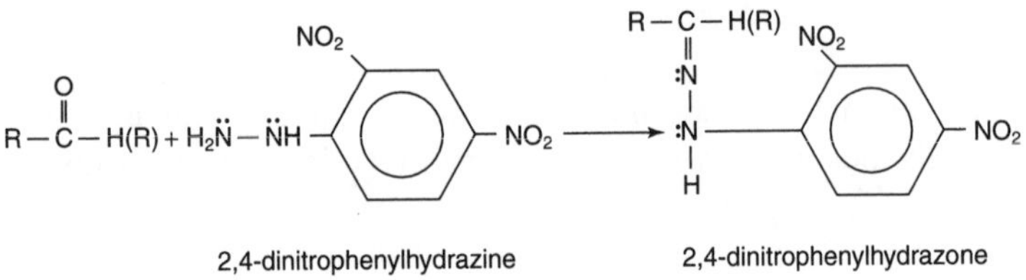

2,4-dinitrophenylhydrazine

2,4-dinitrophenylhydrazone

$$R - \overset{\overset{\displaystyle O}{\|}}{C} - H(R) + H_2 \ddot{N}OH \longrightarrow R - \overset{\overset{\displaystyle \ddot{N}}{|}}{\underset{\overset{\displaystyle |}{OH}}{C}} - H(R)$$

hydroxylamine   oxime

$$R - \overset{\overset{\displaystyle O}{\|}}{C} - H(R) + H_2\ddot{N}\ddot{N}H - \overset{\overset{\displaystyle O}{\|}}{C} - \ddot{N}H_2 \longrightarrow R - \overset{\displaystyle |}{\underset{\displaystyle NH_2}{C}} - H(R)$$

semicarbazide

$$R - C - H(R)$$
$$\overset{|}{\ddot{N}}$$
$$\overset{|}{\ddot{N}} - H$$
$$\overset{|}{C} = O$$
$$\overset{|}{NH_2}$$

semicarbazone

Oximes, 2,4-dinitrophenylhydrazones, and semicarbazones are often used in qualitative organic chemistry as derivatives for aldehydes and ketones.

Oxidations of aldehydes and ketones

Aldehydes can be oxidized to carboxylic acid with both mild and strong oxidizing agents. However, ketones can be oxidized to various types of compounds only by using extremely strong oxidizing agents. Typical oxidizing agents for aldehydes include either potassium permanganate ($KMnO_4$) or potassium dichromate ($K_2Cr_2O_7$) in acid solution and Tollens reagent. Peroxy acids, such as peroxybenzoic acid:

$$(C_6H_5\overset{\overset{\displaystyle O}{\|}}{C} - OOH) \quad \text{are used to oxidize ketones.}$$

benzaldehyde → benzoic acid

cyclohexanecarbaldehyde → cyclohexanoic acid

cyclohexylmethylketone → cyclohexylacetate

Baeyer-Villiger oxidation is a ketone oxidation, and it requires the extremely strong oxidizing agent peroxybenzoic acid. For example, peroxybenzoic acid oxidizes phenyl methyl ketone to phenyl acetate (an ester).

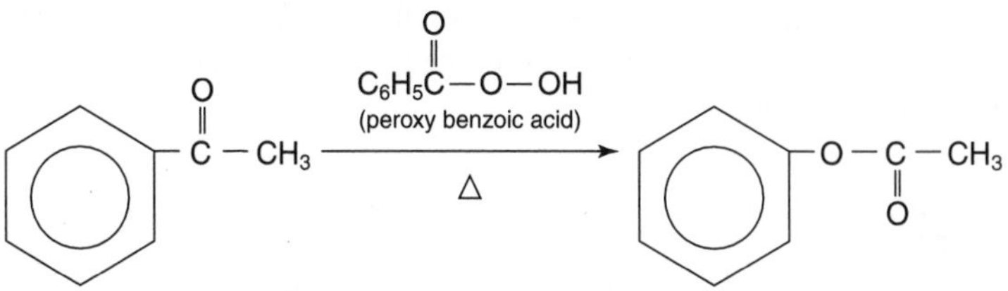

phenyl methyl ketone　　　　　　　　　phenylacetate

Aldol reactions

In addition to nucleophilic additions, aldehydes and ketones show an unusual acidity of hydrogen atoms attached to carbons alpha (adjacent) to the carbonyl group. These hydrogens are referred to as α hydrogens, and the carbon to which they are bonded is an α carbon. In ethanal, there is one α carbon and three α hydrogens, while in acetone there are two α carbons and six α hydrogens.

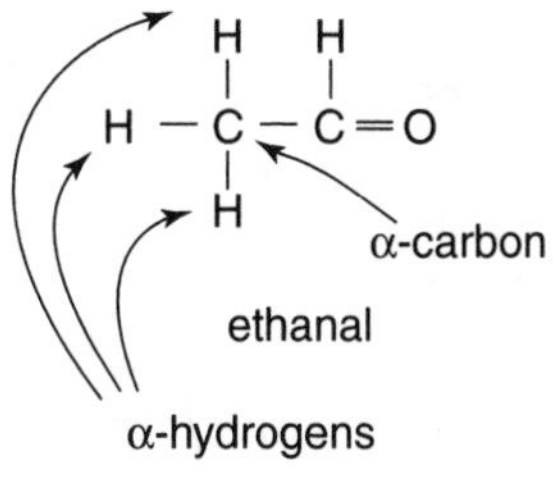

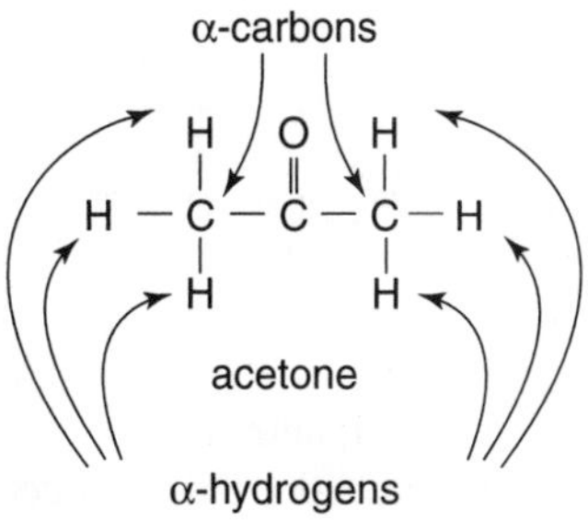

Although weakly acidic (K, 10^{19} to 10^{20}), α hydrogens can react with strong bases to form anions. The unusual acidity of α hydrogens can be explained by both the electron withdrawing ability of the carbony group and resonance in the anion that forms. The electron withdrawing ability of a carbonyl group is caused by the group's dipole nature, which results from the differences in electronegativity between carbon and oxygen.

The anion formed by the loss of an α hydrogen can be resonance stabilized because of the mobility of the π electrons that are on the adjacent carbonyl group.

$$R - \underset{\underset{}{\overset{:\ddot{O}}{\|}}}{C} - \underset{\overset{H}{\underset{H}{|}}}{C} - H \xrightarrow{\ ^{-}OH\ } R - \underset{\overset{\overset{:\ddot{O}:\ -}{\|}}{}}{C} - \underset{\overset{}{\underset{H}{|}}}{\ddot{C}} - H \longleftrightarrow R - \underset{}{C} = \underset{\overset{}{\underset{H}{|}}}{\overset{:\ddot{O}:}{C}} - H$$

keto   enol

The resonance, which stabilizes the anion, creates two resonance structures — an enol and a keto form. In most cases, the keto form is more stable.

Halogenation of ketones

In the presence of a base, ketones with α hydrogens react to form α haloketones.

$$R - \underset{\overset{}{\underset{H}{|}}}{\overset{\overset{H}{|}}{C}} - \underset{\overset{O}{\|}}{C} - R' \xrightarrow[Br_2]{\ ^{-}OH\ } R - \underset{\overset{}{\underset{R}{|}}}{\overset{\overset{Br}{|}}{C}} - \underset{\overset{O}{\|}}{C} - R'$$

Likewise, when methyl ketones react with iodine in the presence of a base, complete halogenation occurs.

$$H - \underset{\overset{}{\underset{H}{|}}}{\overset{\overset{H}{|}}{C}} - \underset{\overset{O}{\|}}{C} - \bigcirc \xrightarrow[I_2]{\ ^{-}OH\ } I - \underset{\overset{}{\underset{I}{|}}}{\overset{\overset{I}{|}}{C}} - \underset{\overset{O}{\|}}{C} - \bigcirc$$

methyl phenyl ketone   triiodomethyl phenyl ketone

The generation of sodium hypoiodate in solution from the reaction of
iodine with sodium hydroxide leads to the formation of iodoform and
sodium benzoate, as shown here.

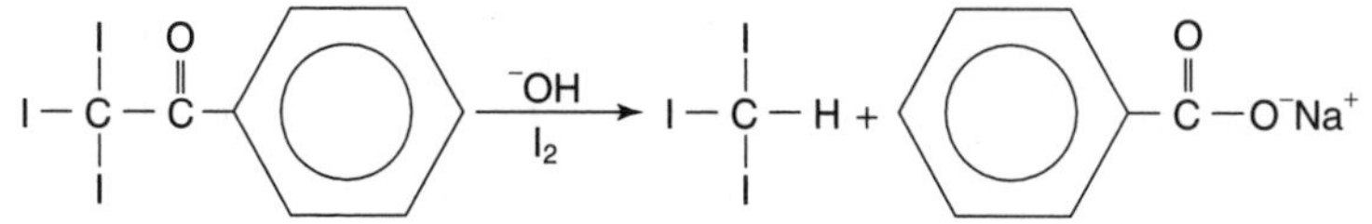

$$I-\underset{I}{\overset{I}{C}}-\overset{O}{\overset{\|}{C}}-\langle\bigcirc\rangle \xrightarrow[\text{I}_2]{^-\text{OH}} I-\underset{I}{\overset{I}{C}}-H + \langle\bigcirc\rangle-\overset{O}{\overset{\|}{C}}-O^-Na^+$$

iodoform   sodium benzoate

Because iodoform is a pale yellow solid, this reaction is often run as
a test for methyl ketones and is called the **iodoform test.**

Aldol condensation

Aldehydes that have α hydrogens react with themselves when mixed
with a dilute aqueous acid or base. The resulting compounds, β-
hydroxy aldehydes, are referred to as **aldol compounds** because they
possess both an aldehyde and alcohol functional group.

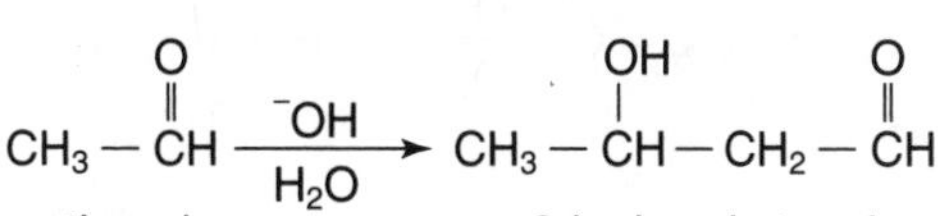

$$CH_3-\overset{O}{\overset{\|}{CH}} \xrightarrow[H_2O]{^-OH} CH_3-\underset{OH}{\overset{}{CH}}-CH_2-\overset{O}{\overset{\|}{CH}}$$

ethanal   3-hydroxybutanal

The aldol condensation proceeds via a carbanion intermediate. The
mechanism of base-catalyzed aldol condensation follows these steps:

1. The base removes an α hydrogen.

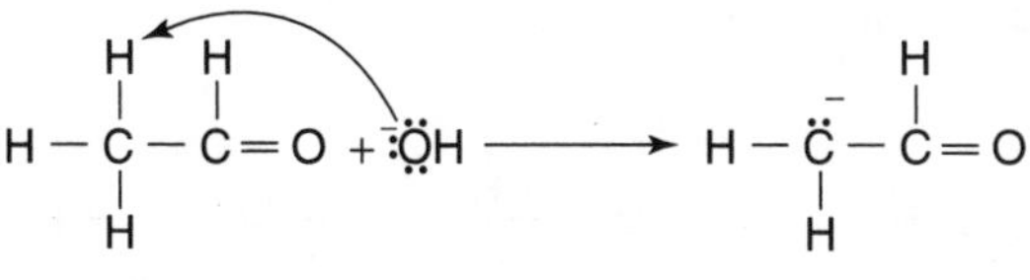

$$H-\underset{H}{\overset{H}{C}}-\overset{H}{C}=O + \ddot{O}H \longrightarrow H-\underset{H}{\overset{\ddot{}}{C}}-\overset{H}{C}=O$$

2. The carbanion undergoes nucleophilic addition with the carbonyl group of a second molecule of ethanal, which leads to formation of the condensation product.

$$H - \overset{\overset{\displaystyle H}{|}}{\underset{\underset{\displaystyle H}{|}}{C}} - \overset{\overset{\displaystyle H}{|}}{C} = \overset{..}{O} + \ :CH_2 - \overset{\overset{\displaystyle H}{|}}{C} = O \longrightarrow H - \overset{\overset{\displaystyle H}{|}}{\underset{\underset{\displaystyle H}{|}}{C}} - \overset{\overset{\displaystyle H}{|}}{\underset{\underset{\displaystyle H}{|}}{C}} - CH_2 - \overset{\overset{\displaystyle H}{|}}{C} = O$$

3. A reaction with water protonates the alkoxide ion.

$$H - \overset{\overset{\displaystyle H}{|}}{\underset{\underset{\displaystyle H}{|}}{C}} - \overset{\overset{\displaystyle :\ddot{O}:^-}{|}}{\underset{\underset{\displaystyle H}{|}}{C}} - CH_2 - \overset{\overset{\displaystyle H}{|}}{C} = O + HOH \longrightarrow H - \overset{\overset{\displaystyle H}{|}}{\underset{\underset{\displaystyle H}{|}}{C}} - \overset{\overset{\displaystyle OH}{|}}{\underset{\underset{\displaystyle H}{|}}{C}} - CH_2 - \overset{\overset{\displaystyle H}{|}}{C} = O + \ ^-OH$$

If the aldol is heated in basic solution, the molecule can be dehydrated to form an α β-unsaturated aldehyde.

$$CH_3 - \overset{\overset{\displaystyle OH}{|}}{\underset{\underset{\displaystyle H}{|}}{C}} - CH_2 - \overset{\overset{\displaystyle H}{|}}{C} = O \xrightarrow[\triangle]{^-OH} CH_3 - \overset{\overset{\displaystyle}{}}{C} = \overset{\overset{\displaystyle}{}}{\underset{\underset{\displaystyle H}{|}}{C}} - \overset{\overset{\displaystyle H}{|}}{\underset{\underset{\displaystyle H}{|}}{C}} = O$$

Cross-aldol condensation

An aldol condensation between two different aldehydes produces a cross-aldol condensation. If both aldehydes possess α hydrogens, a series of products will form. To be useful, a cross-aldol must be run between an aldehyde possessing an α hydrogen and a second aldehyde that does not have α hydrogens.

$$(CH_3)_3CCHO + CH_3CHO \xrightarrow[H_2O]{^-OH} (CH_3)_3CCH = CHCHO$$

Ketonic aldol condensation

Ketones are less reactive towards aldol condensations than aldehydes. With acid catalysts, however, small amounts of aldol product
can be formed. But the Aldol product that forms will rapidly dehy-
drate to form a resonance-stabilized product. This dehydration step
drives the reaction to completion.

2-propanone   4-hydroxy-4-methyl-2-pentanone   4-methyl-3-pentene-2-one

The acid-catalyzed aldol condensation includes two key steps: the
conversion of the ketone into its enolic form, and the attack on a protonated carbonyl group by the enol. The mechanism proceeds as
follows:

1. The oxygen of the carbonyl group is protonated.

2. A water molecule acting as a base removes an acidic α hydrogen, which leads to an enol.

3. The enol attacks a protonated carbonyl group of a second ketone molecule.

$$CH_3CCH_3 + :\overset{..}{O} - H \xrightarrow{H\overset{..}{O}H} CH_3 - C - CH_2 - C - CH_3 + H_3O^+$$

Cyclizations via aldol condensation

Internal aldol condensations (condensations where both carbonyl groups are on the same chain) lead to ring formation.

$$CH_3CCH_2CH_2CH_2CH_2CH_2C - H \xrightarrow{:\overset{..}{O}H}$$

The mechanism for cyclization via an aldol proceeds through an enolate attack on the aldehyde carbonyl.

1. The hydroxy ion removes a hydrogen ion α to the ketone carbonyl.

$$CH_3C - C - CH_2CH_2CH_2CH_2 - C - H + {}^-OH \longrightarrow$$

$$CH_3C - \underset{..}{C} - CH_2CH_2CH_2CH_2 - C - H + H_2O$$

2. The enolate ion attacks the aldehyde carbonyl, closing the ring.

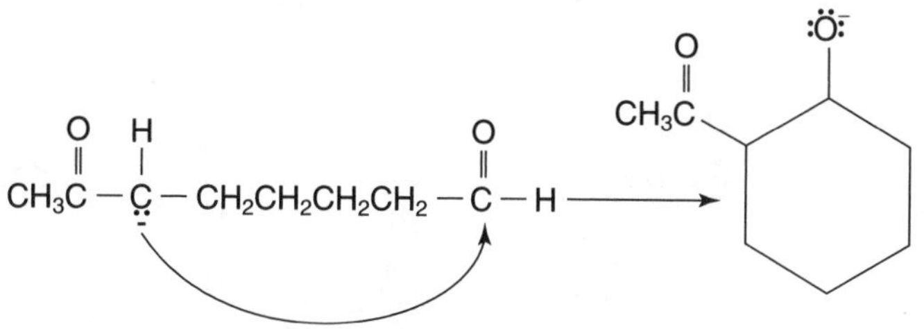

3. The alkoxide ion abstracts a proton from water in an acid-base reaction.

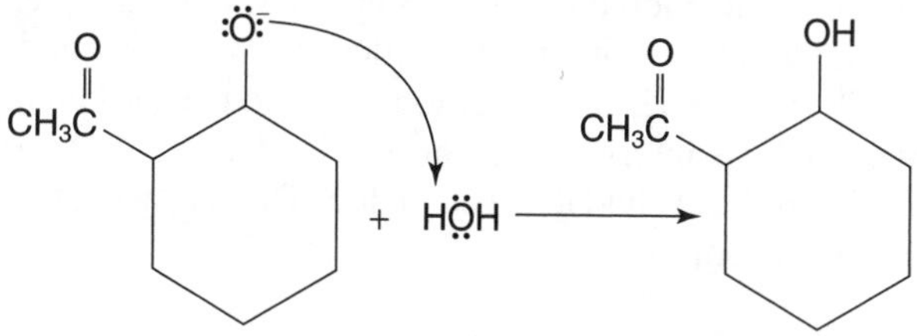

4. The base removes a hydrogen ion to form a resonance-stabilized molecule.

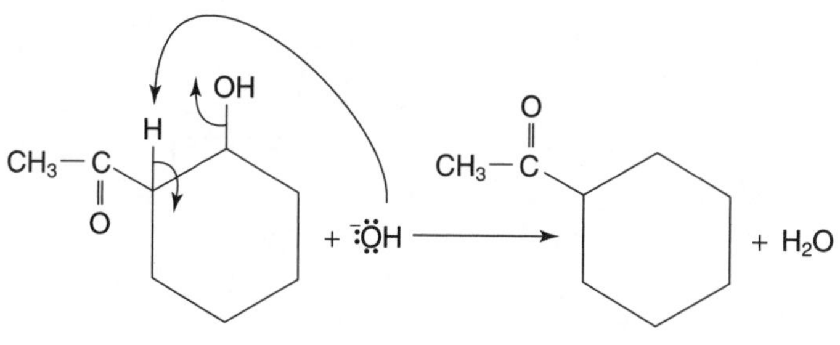

The benzoin condensation

Aromatic aldehydes form a condensation product when heated with a cyanide ion dissolved in an alcohol-water solution. This condensation leads to the formation of α hydroxy ketones.

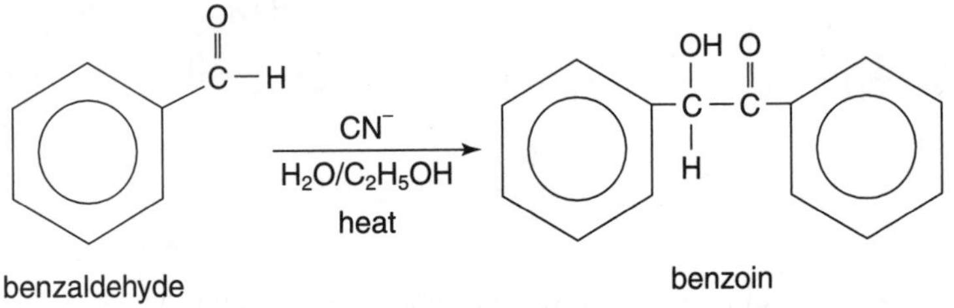

benzaldehyde

benzoin

The cyanide ion is the only known catalyst for this condensation, because the cyanide ion has unique properties. For example, cyanide ions are relatively strong nucleophiles, as well as good leaving groups. Likewise, when a cyanide ion bonds to the carbonyl group of the aldehyde, the intermediate formed is stabilized by resonance between the molecule and the cyanide ion. The following mechanism illustrates these points.

The benzoin condensation reaction proceeds via a nucleophilic substitution followed by a rearrangement reaction.

1. The cyanide ion is attracted to the carbon atom of the carbonyl group.

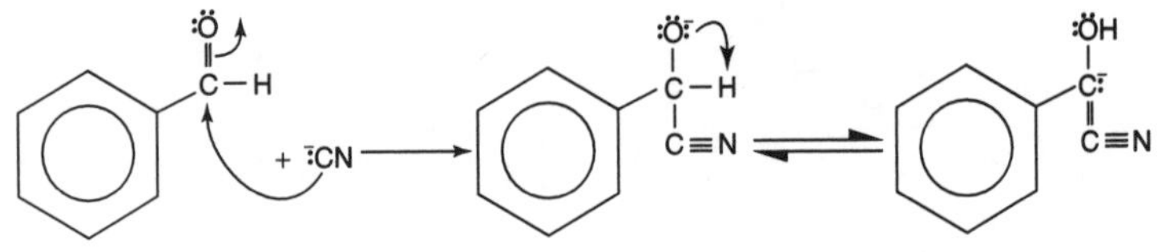

2. The carbanion is resonance-stabilized.

3. The carbanion attacks a second molecule of benzaldehyde.

4. The alkoxide ion removes a proton from the hydroxide group.

5. A pair of electrons on the alkoxide ion are attracted to the carbon bonded to the cyanide group, which then leaves to generate the product.

Introduction

Carboxylic acids are compounds that contain the carboxyl group:

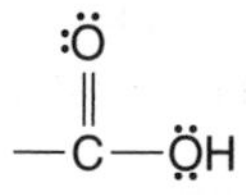

$$\overset{\displaystyle :\!\ddot{O}}{\underset{\displaystyle}{\|}} \\ -C-\ddot{O}H$$

These compounds and their common derivatives make up the bulk of organic compounds. Their common derivatives include acid halides:

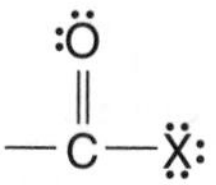

$$\overset{:\!\ddot{O}}{\underset{}{\|}} \\ -C-\ddot{X}\!:$$

acid anhydrides:

$$\overset{:\!\ddot{O}\qquad\ddot{O}\!:}{\underset{}{\|\qquad\|}} \\ -C-O-C-R$$

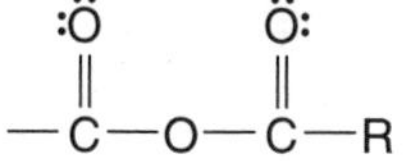

esters:

$$\overset{:\!\ddot{O}}{\underset{}{\|}} \\ -C-\ddot{O}R$$

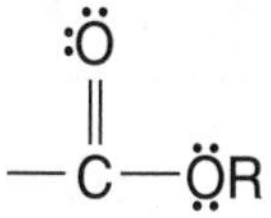

and amides:

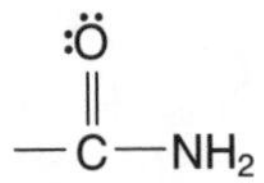

$$
\overset{\displaystyle \ddot{O}}{\underset{\displaystyle}{\underset{\displaystyle}{\overset{\|}{-\,C\,-NH_2}}}}
$$

Nomenclature of carboxylic acids

Two systems are used for naming carboxylic acids: the common system and the IUPAC system.

Common names for carboxylic acids are derived from Latin or Greek words that indicate one of their naturally occurring sources. Table 7-1 lists the common name, structure, source, and etymology for some common carboxylic acids.

Table 7-1: Common Names of Carboxylic Acids

Name	Structure	Source	Etymology
Formic acid	$H-\overset{O}{\overset{\|}{C}}-OH$	Ant	*Formica* (Latin)
Acetic acid	$CH_3-\overset{O}{\overset{\|}{C}}-OH$	Vinegar	*Acetum* (Latin)
Butyric acid	$CH_3-CH_2-CH_2-\overset{O}{\overset{\|}{C}}-OH$	Butter	*Butyrum* (Latin)
Caproic acid	$CH_3-CH_2-CH_2-CH_2-CH_2-\overset{O}{\overset{\|}{C}}-OH$	Goat	*Caper* (Latin)
Steric acid	$CH_3-(CH_2)_{16}-\overset{O}{\overset{\|}{C}}-OH$	Tallow	*Steak* (Greek)

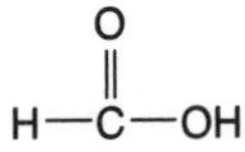

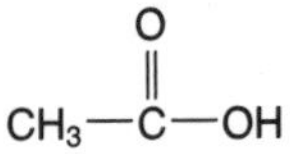

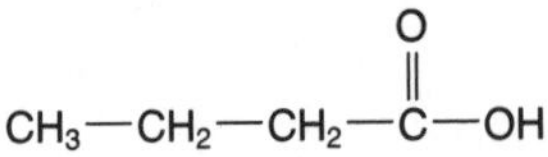

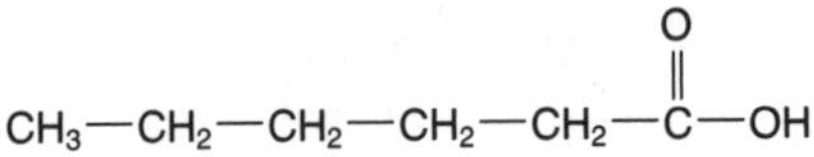

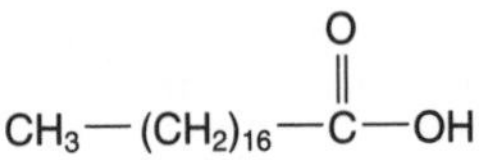

Employ the following steps to derive the IUPAC name for a carboxylic acid:

1. Pick out the longest, continuous chain of carbon atoms that contains the carboxyl group. The parent name for the compound comes from the alkane name for that number of carbon atoms.

2. Change the -e ending of the alkane name to -oic and add the word "acid."

3. Locate and name any substituents, labeling their placement by numbering away from the carboxyl group.

Applying these rules gives the following compound the name 2-ethyl-4-methylpentanoic acid.

$$CH_3-CH-CH_2-CH-C-OH$$

with branches: CH_3 on the second carbon, CH_2-CH_3 (CH_2 bearing CH_3) on the fourth carbon, and O (double bond) on the carboxyl carbon.

Naming carboxylic acid salts

Carboxylic acid salts are named in both the common and IUPAC systems by replacing the -ic ending of the acid name with -ate. For example, $CH_3COO^-K^+$ is potassium acetate or potassium methanoate.

Acidity of carboxylic acids

Carboxylic acids show K_a values in the order of 10^{-4} to 10^{-5} and thus readily react with ordinary aqueous bases such as sodium hydroxide and sodium bicarbonate. This acidity is due to two factors. First, the oxygen atom of the carboxyl group bonded to the hydrogen atom has a partial positive charge on it because of resonance.

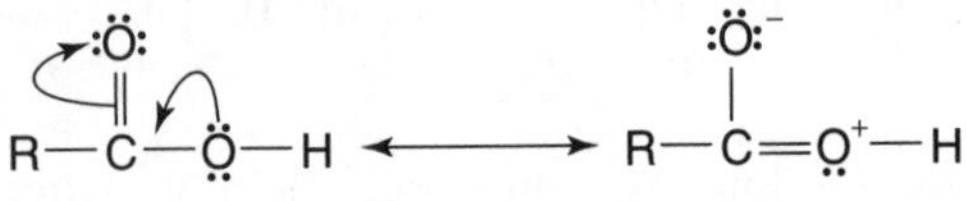

Second, the anion that results from the removal of the hydrogen attached to the carboxyl oxygen is resonance stabilized.

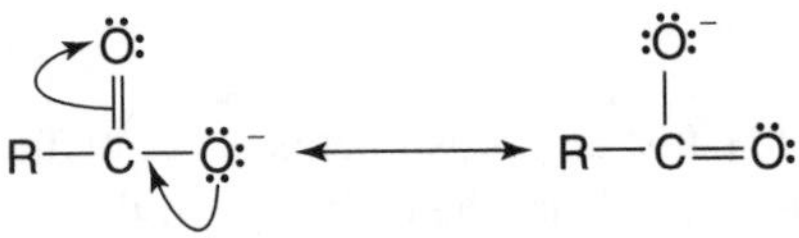

Substituting electron withdrawing groups such as halogens on the chain of the R group(s) increases the acidity of the acid. This effect is strongest for α-substitutions and decreases rapidly as the electron withdrawing group is moved further down the chain.

Preparation of Carboxylic Acids

Carboxylic acids are mainly prepared by the oxidation of a number of different functional groups, as the following sections detail.

Oxidation of alkenes

Alkenes are oxidized to acids by heating them with solutions of potassium permanganate ($KMnO_4$) or potassium dichromate ($K_2Cr_2O_7$).

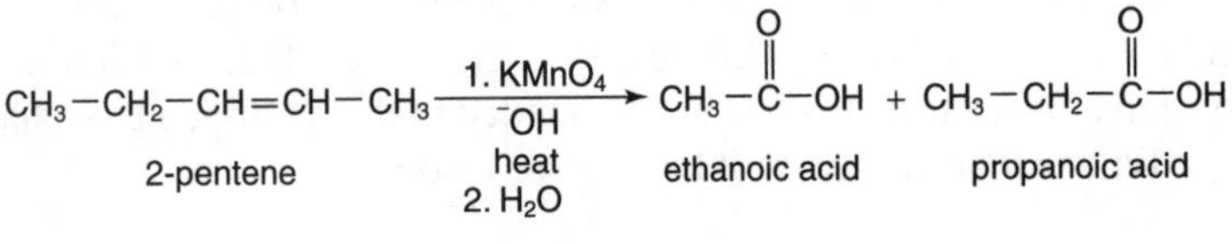

Ozonolysis of alkenes

The **ozonolysis** of alkenes produces aldehydes that can easily be further oxidized to acids.

$$CH_3-CH_2-CH{=}CH-CH_3 \longrightarrow CH_3-\overset{\displaystyle O}{\overset{\|}{C}}-OH + CH_3-CH_2-\overset{\displaystyle O}{\overset{\|}{C}}-OH$$

2-pentene   ethanoic acid   propanoic acid

The oxidation of primary alcohols and aldehydes

The oxidation of primary alcohols leads to the formation of aldehydes that undergo further oxidation to yield acids. All strong oxidizing agents (potassium permanganate, potassium dichromate, and chromium trioxide) can easily oxidize the aldehydes that are formed. **Remember:** Mild oxidizing agents such as manganese dioxide (MnO_2) and Tollen's reagent [$Ag(NH_3)_2{}^+OH^-$] are only strong enough to oxidize alcohols to aldehydes.

$$CH_3-CH_2-OH \xrightarrow[\substack{\text{heat} \\ \text{2. }H_2O}]{\substack{\text{1. }KMnO_4 \\ {}^-OH}} CH_3-\overset{\displaystyle O}{\overset{\|}{C}}-OH$$

ethanol   ethanoic acid

$$CH_3-\overset{\displaystyle O}{\overset{\|}{C}}H \xrightarrow[\substack{\text{warm} \\ \text{2. }H_2O}]{\substack{\text{1. }KMnO_4 \\ H^+}} CH_3-\overset{\displaystyle O}{\overset{\|}{C}}-OH$$

ethanol   ethanoic acid

The oxidation of alkyl benzenes

Alkyl groups that contain **benzylic hydrogens**—hydrogen(s) on a carbon α to a benzene ring—undergo oxidation to acids with strong oxidizing agents.

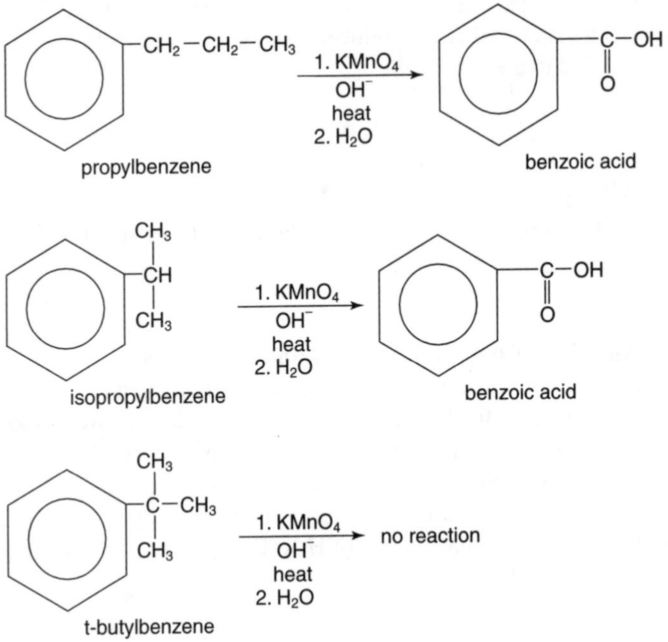

propylbenzene → benzoic acid

isopropylbenzene → benzoic acid

t-butylbenzene → no reaction

In the above example, t-butylbenzene does not contain a benzylic hydrogen and therefore doesn't undergo oxidation.

Hydrolysis of nitriles

The hydrolysis of **nitriles**, which are organic molecules containing a cyano group, leads to carboxylic acid formation. These hydrolysis reactions can take place in either acidic or basic solutions.

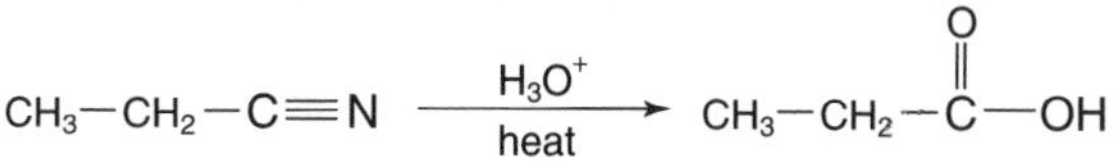

$$CH_3-CH_2-C\equiv N \xrightarrow[\text{heat}]{H_3O^+} CH_3-CH_2-\overset{\displaystyle O}{\overset{\|}{C}}-OH$$

propanenitrile   propanoic acid

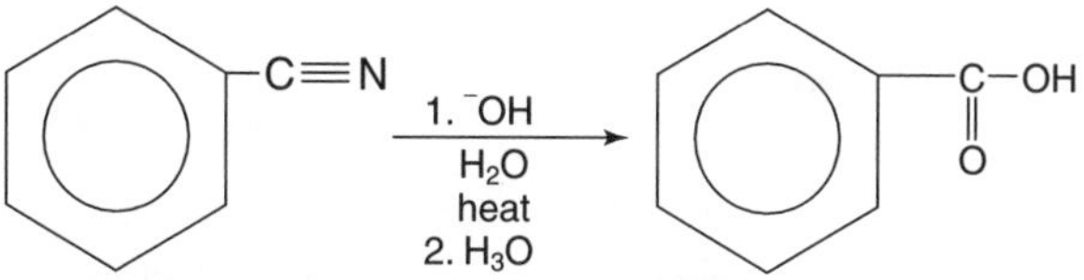

benzonitrile   benzoic acid

The mechanism for these reactions involves the formation of an amide followed by hydrolysis of the amide to the acid. The mechanism follows these steps:

1. The nitrogen atom of the nitrile group is protonated.

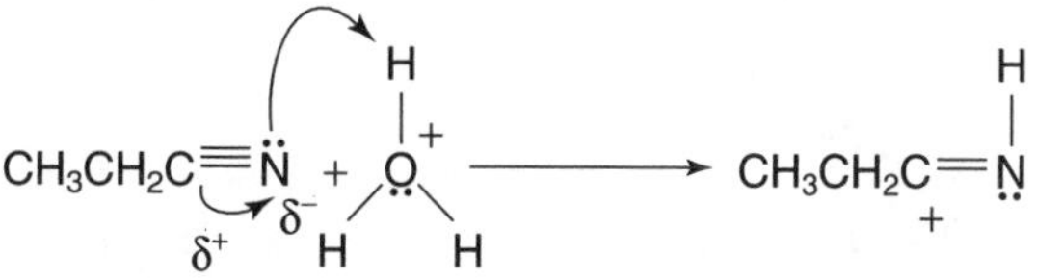

2. The carbocation generated in Step 1 attracts a water molecule.

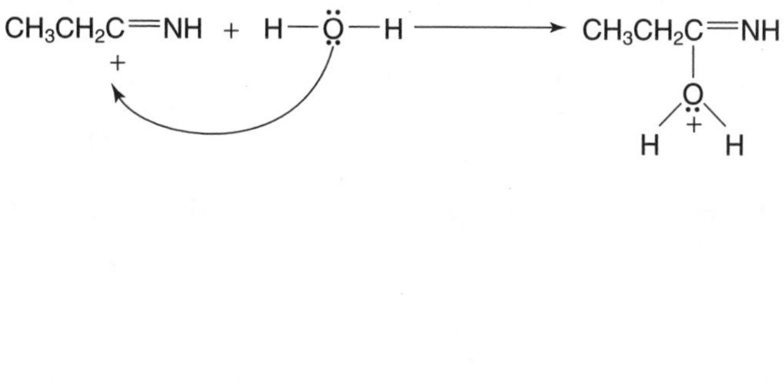

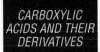

3. The oxonium ion loses a proton to the nitrogen atom, forming an enol.

$$CH_3CH_2C = \overset{\cdot\cdot}{N}H \longrightarrow CH_3CH_2C = \overset{\overset{H}{|}}{N}H$$
$$\underset{\underset{H}{|}}{\overset{\cdot\cdot}{O}} \qquad \qquad \underset{H\overset{\cdot\cdot}{O}:}{+}$$

4. The enol tautomerizes to the more stable keto form.

$$CH_3CH_2C = \overset{\overset{H}{|}}{N}H \rightleftharpoons CH_3 - CH_2 - \overset{\overset{O}{||}}{C} - NH_2 + H^+$$
$$\underset{H\overset{\cdot\cdot}{O}:}{+} \qquad \qquad \text{propanamide}$$

5. The amide is protonated by the acid, forming a carbocation.

$$CH_3 - CH_2 - \overset{\overset{\cdot\cdot}{O}:}{C} - NH_2 + H^+ \longrightarrow CH_3CH_2\overset{+}{C} - NH_2$$
$$\underset{H\overset{\cdot\cdot}{O}:}{}$$

6. A water molecule is attracted to the carbocation.

$$CH_3CH_2\overset{+}{C} - NH_2 + H - \overset{\cdot\cdot}{O} - H \longrightarrow CH_3CH_2\overset{\overset{H \quad H}{\underset{+}{\diagdown} \underset{+}{\diagup}}}{\underset{\overset{\cdot\cdot}{O}}{C}} - NH_2$$
$$\underset{HO}{} \qquad \qquad \qquad \underset{OH}{}$$

7. The oxonium ion loses a proton.

$$\begin{array}{cc} \overset{\displaystyle H \quad H}{\underset{\displaystyle}{\diagdown \overset{+}{\underset{\cdot\cdot}{O}} \diagdown}} & \\ CH_3CH_2\overset{|}{\underset{\displaystyle \underset{\cdot\cdot}{:}OH}{C}}-NH_2 & \longrightarrow \end{array} \qquad CH_3CH_2\overset{\displaystyle :\overset{\cdot\cdot}{O}H}{\underset{\displaystyle :\overset{\cdot\cdot}{O}H}{\overset{|}{\underset{|}{C}}}}-NH_2$$

8. The amine group is protonated.

$$CH_3CH_2-\overset{\displaystyle :\overset{\cdot\cdot}{O}H}{\underset{\displaystyle :\overset{\cdot\cdot}{O}H}{\overset{|}{\underset{|}{C}}}}-\overset{\cdot\cdot}{N}H_2 \quad \xrightarrow{\ H^+\ } \quad CH_3CH_2\overset{\displaystyle :\overset{\cdot\cdot}{O}H}{\underset{\displaystyle :\overset{\cdot\cdot}{O}H}{\overset{|}{\underset{|}{C}}}}-\overset{+}{N}H_3$$

9. An electron pair on one of the oxygens displaces the ammonium group from the molecule.

$$CH_3CH_2\overset{\displaystyle :\overset{\cdot\cdot}{O}H}{\underset{\displaystyle \overset{\cdot\cdot}{C}:OH}{\overset{|}{\underset{|}{C}}}}-\overset{+}{N}H_3 \quad \longrightarrow \quad CH_3CH_2\overset{\displaystyle O}{\overset{\|}{C}}-OH \quad + \quad \overset{\cdot\cdot}{N}H_3$$

The carbonation of Grignard reagents

Grignard reagents react with carbon dioxide to yield acid salts, which, upon acidification, produce carboxylic acids.

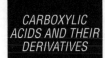
phenyl magnesium bromide   benzoic acid

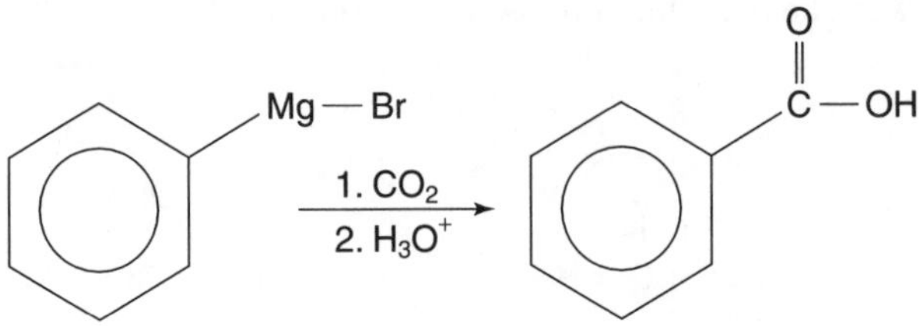

Synthesis of substituted acetic acids via acetoacetic ester
Acetoacetic ester, an ester formed by the self-condensation of ethyl acetate via a Claisen condensation, has the following structure:

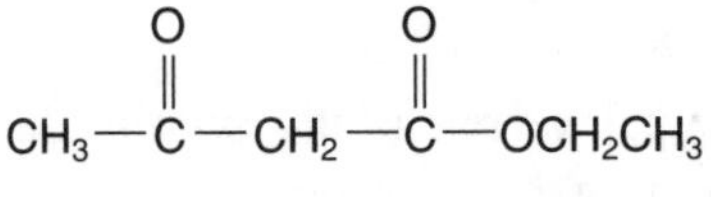

$$CH_3-\overset{O}{\overset{\|}{C}}-CH_2-\overset{O}{\overset{\|}{C}}-OCH_2CH_3$$

The hydrogens on the methylene unit located between the two carbonyl functional groups are acidic due to the electron withdrawing effects of the carbonyl groups. Either or both of these hydrogens can be removed by reaction with strong bases.

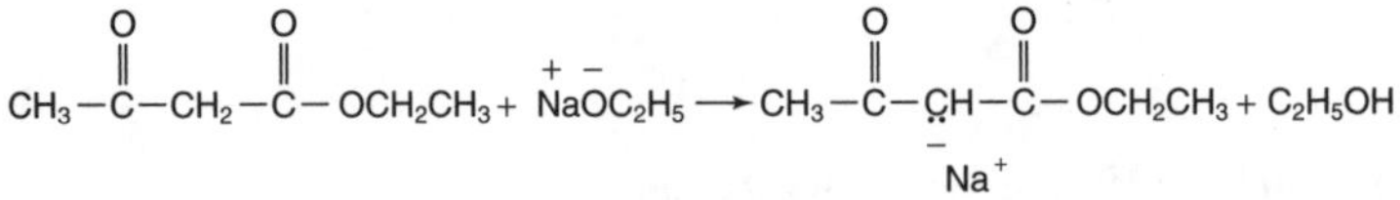

$$CH_3-\overset{O}{\overset{\|}{C}}-CH_2-\overset{O}{\overset{\|}{C}}-OCH_2CH_3 + \overset{+}{Na}\overset{-}{OC_2H_5} \longrightarrow CH_3-\overset{O}{\overset{\|}{C}}-\underset{\underset{Na^+}{-}}{C}H-\overset{O}{\overset{\|}{C}}-OCH_2CH_3 + C_2H_5OH$$

The resulting carbanions can participate in typical S_N reactions that allow the placement of alkyl groups on the chain.

$$CH_3-\overset{\overset{\displaystyle O}{\|}}{C}-\underset{\underset{Na^+}{\bar{\;\;}}}{\overset{\overset{\displaystyle O}{\|}}{CH}}-\overset{\overset{\displaystyle O}{\|}}{C}-OCH_2CH_3 + CH_3Cl \longrightarrow CH_3-\overset{\overset{\displaystyle O}{\|}}{C}-\underset{\underset{CH_3}{|}}{CH}-\overset{\overset{\displaystyle O}{\|}}{C}-OCH_2CH_3 + Na^+Cl^-$$

Hydrolysis of the resulting product with concentrated sodium hydroxide solution liberates the sodium salt of the substituted acid.

$$CH_3-\overset{\overset{\displaystyle O}{\|}}{C}-\underset{\underset{CH_3}{|}}{CH}-\overset{\overset{\displaystyle O}{\|}}{C}-OCH_2CH_3 + NaOH \xrightarrow[\text{heat}]{H_2O} CH_3-\overset{\overset{\displaystyle O}{\|}}{C}-\bar{O}Na^+ + CH_3CH_2\overset{\overset{\displaystyle O}{\|}}{C}-\bar{O}Na^+ + CH_3CH_2OH$$

Addition of aqueous acid liberates the substituted acid.

$$CH_3CH_2\overset{\overset{\displaystyle O}{\|}}{C}-O^-Na^+ + CH_3\overset{\overset{\displaystyle O}{\|}}{C}-O^-Na^+ \xrightarrow[H_2O]{HCl} CH_3CH_2\overset{\overset{\displaystyle O}{\|}}{C}-OH + CH_3\overset{\overset{\displaystyle O}{\|}}{C}-OH + NaCl$$

The second hydrogen on the methylene unit of acetoacetic ester can also be replaced by an alkyl group, creating a disubstituted acid. To accomplish this conversion, the reaction product in step 2 above would be reacted with a very strong base to create a carbanion.

$$CH_3-\overset{\overset{\displaystyle O}{\|}}{C}-\underset{\underset{CH_3}{|}}{CH}-\overset{\overset{\displaystyle O}{\|}}{C}-OCH_2CH_3 + (CH_3)_3CO^- K^+ \longrightarrow CH_3-\overset{\overset{\displaystyle O}{\|}}{C}-\underset{\underset{CH_3}{|}}{\overset{\overset{\displaystyle K^+}{\bar{\;\;}}}{C}}-\overset{\overset{\displaystyle O}{\|}}{C}-OCH_2CH_3 + (CH_3)_3COH$$

This carbanion can participate in a typical S_N reaction, allowing the placement of a second alkyl group on the chain.

$$CH_3-\overset{\overset{\displaystyle O}{\|}}{C}-\overset{\overset{\displaystyle K^+}{\underset{\underset{\displaystyle CH_3}{|}}{\overset{-}{C}}}}{}-\overset{\overset{\displaystyle O}{\|}}{C}-OCH_2CH_3 + CH_3CH_2Cl \longrightarrow CH_3\overset{\overset{\displaystyle O}{\|}}{C}-\overset{\overset{\displaystyle CH_2CH_3}{|}}{\underset{\underset{\displaystyle CH_3}{|}}{C}}-\overset{\overset{\displaystyle}{}}{\underset{\underset{\displaystyle O}{\|}}{C}}-OCH_2CH_3 + KCl$$

Hydrolysis using concentrated aqueous sodium hydroxide leads to the formation of the sodium salt of the disubstituted acid.

$$CH_3\overset{\overset{\displaystyle O}{\|}}{C}-\overset{\overset{\displaystyle CH_2CH_3}{|}}{\underset{\underset{\displaystyle CH_3}{|}}{C}}-\overset{\overset{\displaystyle}{}}{\underset{\underset{\displaystyle O}{\|}}{C}}-OCH_2CH_3 + NaOH \xrightarrow[\text{heat}]{H_2O} CH_3\overset{\overset{\displaystyle O}{\|}}{C}-O^-Na^+ + CH_3\overset{\underset{\underset{\displaystyle CH_2CH_3}{|}}{}}{CH}-\overset{\overset{\displaystyle O}{\|}}{C}-O^-Na^+ + CH_3CH_2OH$$

Addition of aqueous acid liberates the disubstituted acid.

$$CH_3\overset{\underset{\underset{\displaystyle CH_2CH_3}{|}}{}}{CH}-\overset{\overset{\displaystyle O}{\|}}{C}-O^-Na^+ \xrightarrow[H_2O]{HCl} CH_3\overset{\underset{\underset{\displaystyle CH_3}{|}}{}}{CH}-\overset{\overset{\displaystyle O}{\|}}{C}-OH + NaCl$$

The acid formed has a methyl and an ethyl group in place of two hydrogens of acetic acid and is therefore often referred to as a disubstituted acetic acid.

If dilute sodium hydroxide were used instead of concentrated, the product formed would be a methyl ketone. This ketone occurs because dilute sodium hydroxide has sufficient strength to hydrolyze the ester functional group but insufficient strength to hydrolyze the ketone functional group. Concentrated sodium hydroxide is strong enough to hydrolyze both the ester functional group and the ketone functional group and, therefore, forms the substituted acid rather than the ketone.

A reaction between a disubstituted acetoacetic ester and dilute sodium hydroxide forms the following products:

$$CH_3\overset{\displaystyle O}{\overset{\|}{C}}-\underset{\underset{\displaystyle CH_3}{|}}{\overset{\overset{\displaystyle CH_2CH_3}{|}}{C}}-\overset{\displaystyle O}{\underset{\|}{C}}-OCH_2CH_3 + \text{dilute NaOH} \longrightarrow CH_3\overset{\displaystyle O}{\overset{\|}{C}}-\underset{\underset{\displaystyle CH_3}{|}}{\overset{\overset{\displaystyle CH_2CH_3}{|}}{C}}-COOH + CH_3CH_2OH$$

β-ketoacid

Upon heating, the β ketoacid becomes unstable and decarboxylates, leading to the formation of the methyl ketone.

$$CH_3\overset{\displaystyle O}{\overset{\|}{C}}-\underset{\underset{\displaystyle CH_3}{|}}{\overset{\overset{\displaystyle CH_2CH_3}{|}}{C}}-\overset{\displaystyle O}{\underset{\|}{C}}-OH \xrightarrow{\;100°\;} CH_3\overset{\displaystyle O}{\overset{\|}{C}}-\underset{\underset{\displaystyle CH_3}{|}}{\overset{\overset{\displaystyle CH_2CH_3}{|}}{C}}-H + CO_2$$

A Claisen condensation of ethyl acetate prepares acetoacetic ester.

$$CH_3-\overset{\displaystyle O}{\overset{\|}{C}}-O-CH_2-CH_3 + \text{NaOH} \longrightarrow CH_3-\overset{\displaystyle O}{\overset{\|}{C}}-CH_2-\overset{\displaystyle O}{\overset{\|}{C}}-OCH_2CH_3 + H_2O$$

The Claisen condensation reaction occurs by a nucleophilic addition to an ester carboxyl group, which follows these steps:

1. An α hydrogen on the ester is removed by a base, which leads to the formation of a carbanion that is resonance stabilized.

$$H-\underset{\underset{H}{|}}{\overset{\overset{H}{|}}{C}}-\overset{\overset{O}{\|}}{C}-OCH_2CH_3 + :\overset{\bar{}}{O}H \longrightarrow H-\underset{\underset{\cdot}{|}}{\overset{\overset{H}{|}}{\underset{\cdot}{C}}}-\overset{\overset{:\ddot{O}}{\|}}{C}-\ddot{O}CH_2CH_3 + H_2O$$

2. Acting as a nucleophile, the carbanion attacks the carboxyl carbon of a second molecule of ester.

$$H-\underset{\underset{H}{|}}{\overset{\overset{H}{|}}{C}}-\overset{\overset{:\ddot{O}}{\|}}{C}-OCH_2CH_3 + H-\underset{\underset{\cdot}{|}}{\overset{\overset{H}{|}}{\underset{\cdot}{C}}}-\overset{\overset{O}{\|}}{C}-\ddot{O}CH_2CH_3 \longrightarrow H-\underset{\underset{H}{|}\ :\ddot{O}CH_2CH_3}{\overset{\overset{H}{|}}{C}}-\underset{}{\overset{\overset{:\ddot{O}:^-}{|}}{C}}-CH_2\overset{\overset{:\ddot{O}}{\|}}{C}-OCH_2CH_3$$

3. A pair of unshared electrons on the alkoxide oxygen move toward the carboxyl carbon, helping the ethoxy group to leave.

$$H-\underset{\underset{H}{|}\ :\ddot{O}CH_2CH_3}{\overset{\overset{H}{|}}{C}}-\overset{\overset{:\ddot{O}:}{|}}{C}-CH_2\overset{\overset{:\ddot{O}}{\|}}{C}-OCH_2CH_3 \longrightarrow H-\underset{\underset{H}{|}}{\overset{\overset{H}{|}}{C}}-\overset{\overset{:\ddot{O}}{\|}}{C}-CH_2\overset{\overset{:\ddot{O}}{\|}}{C}-\ddot{O}CH_2CH_3 + CH_3CH_2\ddot{O}:^-$$

Synthesis of substituted acetic acid via malonic ester

Malonic ester is an ester formed by reacting an alcohol with malonic acid (propanedicarboxylic acid). Following is the structure of diethyl malonate:

$$CH_3-CH_2-O-\overset{\overset{O}{\|}}{C}-CH_2-\overset{\overset{O}{\|}}{C}-O-CH_2-CH_3$$

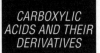

The hydrogen atoms on the methylene unit between the two carboxyl groups are acidic like those in acetoacetic ester. Strong bases can remove these acidic hydrogens.

$$CH_3-CH_2-O-\overset{\overset{O}{\|}}{C}-CH_2-\overset{\overset{O}{\|}}{C}-O-CH_2-CH_3 \ + \ C_2H_5O^-Na^+ \longrightarrow$$

$$CH_3-CH_2-O-\overset{\overset{O}{\|}}{C}-\underset{Na^+}{\underset{|}{CH}}-\overset{\overset{O}{\|}}{C}-O-CH_2-CH_3 \ + \ C_2H_5OH$$

The resulting carbocation can participate in typical S_N reactions, allowing the placement of an alkyl group on the chain.

$$CH_3-CH_2-O-\overset{\overset{O}{\|}}{C}-\underset{Na^+}{CH}-\overset{\overset{O}{\|}}{C}-O-CH_2-CH_3 \ + \ CH_3Br \longrightarrow$$

$$CH_3-CH_2-O-\overset{\overset{O}{\|}}{C}-\underset{\underset{CH_3}{|}}{CH}-\overset{\overset{O}{\|}}{C}-O-CH_2-CH_3 \ + \ NaBr$$

A second alkyl group can be placed on the compound by reacting the product formed in the previous step with a very strong base to form a new carbanion.

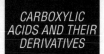

$$CH_3-CH_2-O-\overset{\overset{O}{\|}}{C}-\underset{\underset{CH_3}{|}}{CH}-\overset{\overset{O}{\|}}{C}-O-CH_2-CH_3 \ + \ (CH_3)_3CO^-K^+ \longrightarrow$$

$$CH_3-CH_2-O-\overset{\overset{O}{\|}}{C}-\underset{\underset{CH_3}{|}}{\overset{K^+}{\underset{-}{CH}}}-\overset{\overset{O}{\|}}{C}-O-CH_2-CH_3$$

The resulting carbanion can participate in a typical S_N reaction, allowing the placement of a second alkyl group on the chain.

$$CH_3-CH_2-O-\overset{\overset{O}{\|}}{C}-\underset{\underset{CH_3}{|}}{\overset{K^+}{\underset{-}{C}}}-\overset{\overset{O}{\|}}{C}-O-CH_2-CH_3 \ + \ CH_3CH_2Br \longrightarrow$$

$$CH_3-CH_2-O-\overset{\overset{O}{\|}}{C}-\underset{\underset{CH_3}{|}}{\overset{CH_2CH_3}{\overset{|}{C}}}-\overset{\overset{O}{\|}}{C}-O-CH_2-CH_3 \ + \ KBr$$

Hydrolysis of the resulting product with concentrated aqueous sodium hydroxide produces the sodium salt of the disubstituted acid.

$$CH_3-CH_2-O-\overset{\overset{O}{\|}}{C}-\underset{\underset{CH_3}{|}}{\overset{CH_2CH_3}{\overset{|}{C}}}-\overset{\overset{O}{\|}}{C}-O-CH_2-CH_3 \ + \ NaOH \xrightarrow{H_2O} Na^+{}^-O-\overset{\overset{O}{\|}}{C}-\underset{\underset{CH_3}{|}}{\overset{CH_2CH_3}{\overset{|}{C}}}-\overset{\overset{O}{\|}}{C}-O^-Na^+$$

Addition of aqueous acid converts the salt into its conjugate acid.

$$Na^+{}^-O-\overset{\overset{O}{\|}}{C}-\underset{\underset{CH_3}{|}}{\overset{\overset{CH_2CH_3}{|}}{C}}-\overset{\overset{O}{\|}}{C}-O^-Na^+ \xrightarrow{H_3O^+} HO-\overset{\overset{O}{\|}}{C}-\underset{\underset{CH_3}{|}}{\overset{\overset{CH_2CH_3}{|}}{C}}-\overset{\overset{O}{\|}}{C}-OH$$

Upon heating, the β ketoacid becomes unstable and decarboxylates, forming a disubstituted acetic acid.

$$HO-\overset{\overset{O}{\|}}{C}-\underset{\underset{CH_3}{|}}{\overset{\overset{CH_2CH_3}{|}}{C}}-\overset{\overset{O}{\|}}{C}-OH \xrightarrow{heat} CH_3CH_2-\underset{\underset{CH_3}{|}}{\overset{\overset{H}{|}}{C}}-\overset{\overset{O}{\|}}{C}-OH \;+\; CO_2$$

α halo acids, α hydroxy acids, and α, β unsaturated acids

The reaction of aliphatic carboxylic acids with bromine in the presence of phosphorous produces α halo acids. This reaction is the **Hell-Volhard-Zelinski reaction.**

$$CH_3CH_2COOH \xrightarrow[\substack{P \\ 2.\,H_2O}]{1.\,Br_2} CH_3\underset{\underset{Br}{|}}{CH}-COOH$$

α halo acids can be converted to α hydroxy acids by hydrolysis.

$$CH_3\underset{\underset{Br}{|}}{CH}-COOH \xrightarrow[2.\,H_3O^+]{1.\,OH^-} CH_3\underset{\underset{OH}{|}}{CH}-COOH$$

α halo acids can be converted to α amino acids by reacting with ammonia.

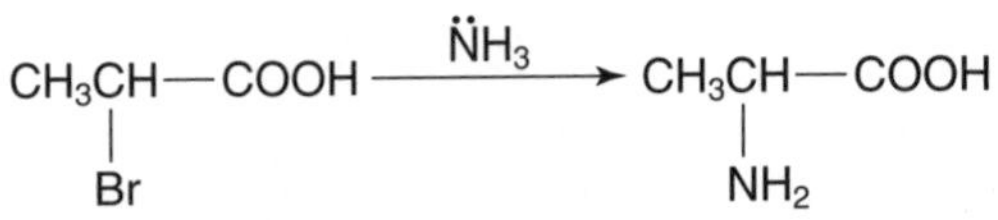

$$CH_3CH-COOH \xrightarrow{\ddot{N}H_3} CH_3CH-COOH$$
$$\quad\;|\qquad\qquad\qquad\qquad\quad\;\;| $$
$$\quad Br \qquad\qquad\qquad\qquad\quad NH_2$$

α halo acids and α hydroxy acids can be converted to α, β unsaturated acids by dehydrohalogenation and dehydration, respectively.

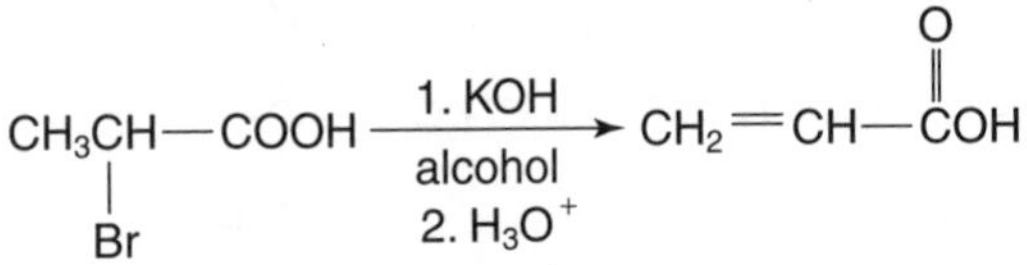

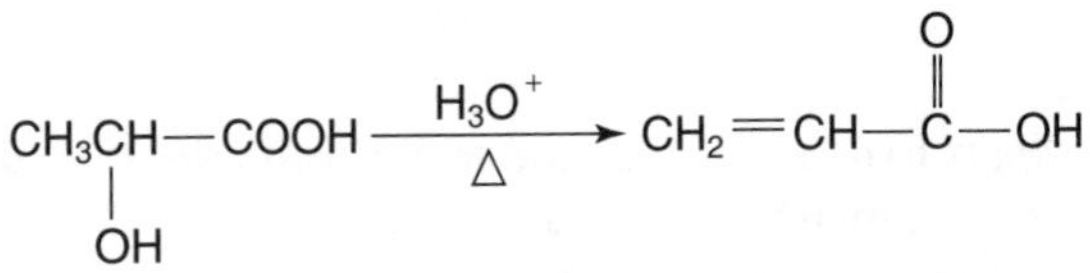

$$CH_3CH-COOH \xrightarrow[\substack{\text{alcohol} \\ 2.\,H_3O^+}]{1.\,KOH} CH_2{=}CH-\overset{\displaystyle O}{\overset{\|}{C}}OH$$
$$\quad\;|$$
$$\quad Br$$

$$CH_3CH-COOH \xrightarrow[\Delta]{H_3O^+} CH_2{=}CH-\overset{\displaystyle O}{\overset{\|}{C}}-OH$$
$$\quad\;|$$
$$\quad OH$$

Reactions of Carboxylic Acids

Carboxylic acids undergo reactions to produce derivatives of the acid. The most common derivatives formed are esters, acid halides, acid anhydrides, and amides.

Ester formation

Esters are compounds formed by the reaction of carboxylic acids with alcohols, and they have a general structural formula of:

$$
\begin{array}{c}
O \\
\parallel \\
R-C-OR'
\end{array}
$$

The simplest method of preparation is the **Fischer method,** in which an alcohol and an acid are reacted in an acidic medium. The reaction exists in an equilibrium condition and does not go to completion unless a product is removed as fast as it forms.

$$
\underset{\text{ethanoic acid}}{CH_3-\overset{\overset{\textstyle O}{\parallel}}{C}-OH} + \underset{\text{ethanol}}{CH_3-CH_2-OH} \rightleftharpoons \underset{\substack{\text{ethylethanoate} \\ \text{(ethyl acetate)}}}{CH_3CH_2\overset{\overset{\textstyle O}{\parallel}}{C}-CH_3}
$$

The Fischer esterification proceeds via a carbocation mechanism. In this mechanism, an alcohol is added to a carboxylic acid by the following steps:

1. The carboxyl carbon of the carboxylic acid is protonated.

$$
CH_3-\overset{\overset{\textstyle :\ddot{O}}{\parallel}}{C}-OH + H^+ \rightleftharpoons CH_3-\overset{\overset{\textstyle :\ddot{O}H}{|}}{\underset{+}{C}}-\ddot{O}H
$$

2. An alcohol molecule adds to the carbocation produced in Step 1.

$$
CH_3-\overset{\overset{\textstyle OH}{|}}{\underset{+}{C}}-OH + CH_3\ddot{O}H \rightleftharpoons CH_3-\overset{\overset{\textstyle OH}{|}}{\underset{\underset{H \;+\; CH_3}{\overset{..}{O}}}{C}}-OH
$$

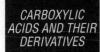

3. A proton is lost from the oxonium ion generated in Step 2.

$$CH_3-\underset{\underset{\underset{H\ +\ CH_3}{}}{\overset{\displaystyle :\ddot{O}}{|}}}{\overset{\displaystyle OH}{\underset{|}{C}}}-OH \;\rightleftharpoons\; CH_3-\underset{\underset{:\ddot{O}-CH_3}{|}}{\overset{\displaystyle OH}{\underset{|}{C}}}-OH \;+\; H^+$$

4. A proton is picked up from solution by a hydroxyl group.

$$CH_3-\underset{\underset{:\ddot{O}CH_3}{|}}{\overset{\displaystyle :\ddot{O}H}{\underset{|}{C}}}-\ddot{O}H \;+\; H^+ \;\rightleftharpoons\; CH_3-\underset{\underset{:\ddot{O}-CH_3}{|}}{\overset{\displaystyle \overset{H\diagdown\ \overset{+}{\ddot{O}}\diagup H}{|}}{\underset{|}{C}}}-\ddot{O}H$$

5. A pair of unshared electrons from the remaining hydroxyl group helps the water molecule leave.

$$CH_3-\underset{\underset{:\ddot{O}CH_3}{|}}{\overset{\displaystyle \overset{H\diagdown\ \overset{+}{\ddot{O}}\diagup H}{|}}{\underset{|}{C}}}-\ddot{O}H \;\rightleftharpoons\; CH_3-\underset{\underset{:\ddot{O}CH_3}{|}}{\overset{\displaystyle }{C}}=\overset{+}{O}H$$

6. The oxonium ion loses a proton to generate the ester.

$$CH_3-\underset{\underset{:\ddot{O}CH_3}{|}}{\overset{\displaystyle }{C}}=\overset{+}{\underset{}{O}}-H \;\rightleftharpoons\; CH_3-\underset{}{\overset{\displaystyle :\ddot{O}}{\underset{\|}{C}}}-\ddot{O}CH_3$$

Nonreversible ester formation

Esters can also be prepared in a nonreversible reaction of an acid with an alkoxide ion.

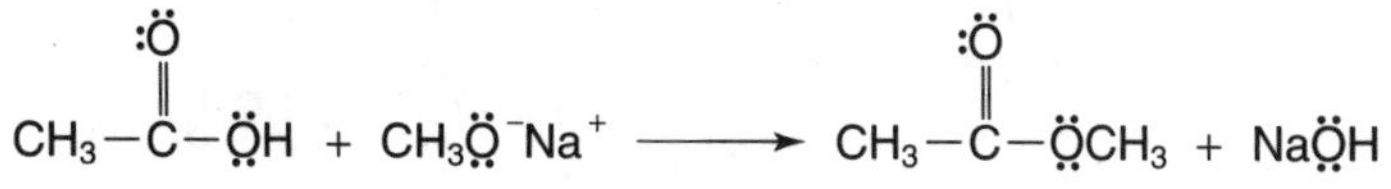

$$CH_3-\overset{\overset{\displaystyle ::\ddot{O}}{\|}}{C}-\ddot{O}H + CH_3\ddot{O}^-Na^+ \longrightarrow CH_3-\overset{\overset{\displaystyle ::\ddot{O}}{\|}}{C}-\ddot{O}CH_3 + Na\ddot{O}H$$

The nonreversible esterification reaction proceeds via a nucleophilic substitution reaction.

1. Acting as a nucleophile, the alkoxide ion is attracted to the carbon atom of the carboxyl group.

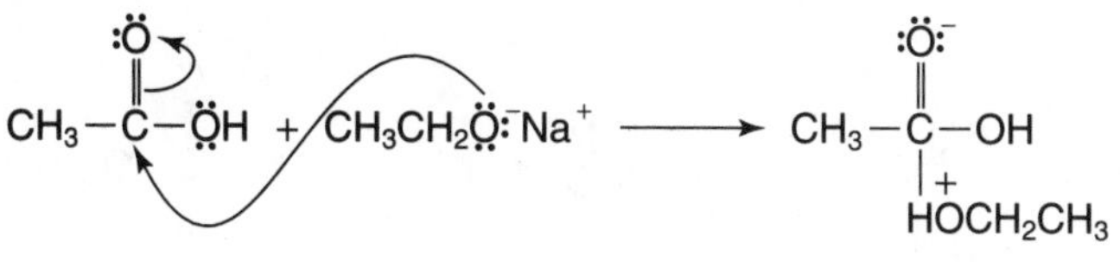

$$CH_3-\overset{\overset{\displaystyle ::\ddot{O}}{\|}}{C}-\ddot{O}H + CH_3CH_2\ddot{O}^-Na^+ \longrightarrow CH_3-\overset{\overset{\displaystyle ::\ddot{O}^-}{|}}{\underset{\overset{|}{HOCH_2CH_3}}{C}}-OH$$

2. The oxonium loses a proton.

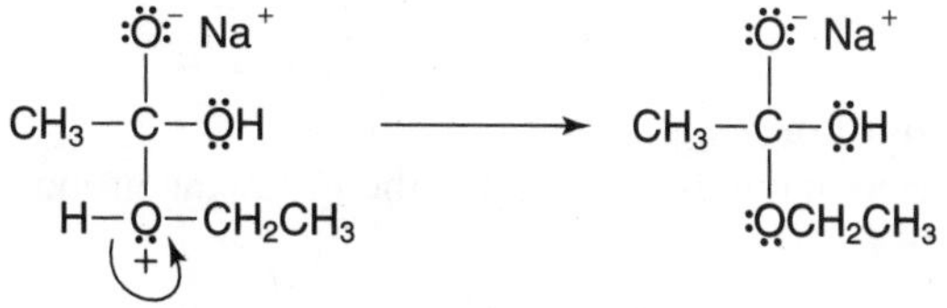

$$CH_3-\overset{\overset{\displaystyle ::\ddot{O}^-\ Na^+}{|}}{\underset{\overset{|}{H-\overset{+}{\underset{}{O}}-CH_2CH_3}}{C}}-\ddot{O}H \longrightarrow CH_3-\overset{\overset{\displaystyle ::\ddot{O}^-\ Na^+}{|}}{\underset{\overset{|}{::\ddot{O}CH_2CH_3}}{C}}-\ddot{O}H$$

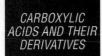
3. An unshared electron pair from the alkoxide ion moves toward
   the carbonyl carbon, assisting the hydroxyl group's exit.

$$CH_3-C-\overset{\displaystyle :\ddot{O}:^{-}\ Na^{+}}{\underset{\displaystyle \overset{|}{O}CH_2CH_3}{\overset{|}{\ddot{O}}H}} \longrightarrow CH_3-\overset{\displaystyle O}{\overset{\|}{C}}-OCH_2CH_3$$

Methyl ester formation

Methyl esters are often prepared by the reaction of carboxylic acids
with diazomethane.

$$\underset{\text{benzoic acid}}{C_6H_5-\overset{\displaystyle O}{\overset{\|}{C}}-OH} \xrightarrow[\text{ether}]{CH_2N_2} \underset{\text{methyl benzoate}}{C_6H_5-\overset{\displaystyle O}{\overset{\|}{C}}-O-CH_3}$$

Amide formation

Amides are compounds that contain the following group:

$$-\overset{\displaystyle O}{\overset{\|}{C}}-NH_2$$

Substituted amides can contain the following groups:

$$\overset{\displaystyle O}{\underset{\displaystyle \|}{}} \qquad \overset{\displaystyle O}{\underset{\displaystyle \|}{}}$$

$$-C-NHR \quad \text{or} \quad -C-R_1R_2$$

An amide name is based on the name of the carboxylic acid of the same number of carbon atoms, but the *-oic* ending is changed to *amide*. Amides with alkyl groups on the nitrogen are substituted amides and are named the same as N-substituted amides, except the parent name is preceded by the name of the alkyl substituent and a capital N precedes the substituent name.

$$CH_3-CH_2-\overset{\overset{\displaystyle O}{\|}}{C}-NH_2 \qquad\qquad CH_3-CH_2-\overset{\overset{\displaystyle O}{\|}}{C}-NH-CH_3$$

propanamide   N-methylpropanamide

$$CH_3-CH_2-\overset{\overset{\displaystyle O}{\|}}{C}-\overset{\overset{\displaystyle CH_3}{|}}{N}-CH_3$$

N,N-dimethylpropanamide

Amides are ordinarily prepared by a reaction of acid chlorides with ammonia or amines.

An amide is prepared by reacting an acid halide with ammonia.

$$CH_3-CH_2-\overset{\overset{\displaystyle O}{\|}}{C}-Cl \xrightarrow[\Delta]{\ddot{N}H_3} CH_3-CH_2-\overset{\overset{\displaystyle O}{\|}}{C}-NH_2 + NH_4Cl$$

An N-substituted amide is prepared by reacting an acid halide with a primary amine.

$$CH_3-CH_2-\overset{\displaystyle O}{\overset{\|}{C}}-Cl \xrightarrow[\Delta]{CH_3\ddot{N}H_2} CH_3-CH_2-\overset{\displaystyle O}{\overset{\|}{C}}-NHCH_3 + NH_4Cl$$

An N,N-disubstituted amide is prepared by reacting an acid halide with a secondary amine.

$$CH_3-CH_2-\overset{\displaystyle O}{\overset{\|}{C}}-Cl \xrightarrow[\Delta]{(CH_3)_2\ddot{N}H} CH_3-CH_2-\overset{\displaystyle O}{\overset{\|}{C}}-N(CH_3)_2 + NH_4Cl$$

You can also react ammonia with esters to prepare primary amides.

$$CH_3-CH_2-\overset{\displaystyle O}{\overset{\|}{C}}-OCH_2CH_3 \xrightarrow[\Delta]{\ddot{N}H_3} CH_3-CH_2-\overset{\displaystyle O}{\overset{\|}{C}}-NH_2 + CH_3CH_2OH$$

The mechanism for amide formation proceeds via attack by the ammonia molecule, which acts as a nucleophile, on the carboxyl carbon of the acid chloride or ester. The alkoxide ion that forms assists with the displacement of the chloride ion or alkoxy group.

1. The ammonia molecule attacks the carboxyl carbon, which leads to the formation of an alkoxide ion.

$$R-\overset{\displaystyle :\ddot{O}}{\overset{\|}{C}}-\ddot{O}R(X) + \underset{\cdot}{N}H_3 \longrightarrow R-\overset{\displaystyle :\ddot{O}:^-}{\underset{\underset{\displaystyle +NH_3}{|}}{\overset{|}{C}}}-\ddot{O}R(X)$$

2. The ammonium ion loses a proton to form an —NH$_2$ group.

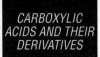

3. An unshared electron pair on the alkoxide ion oxygen moves in to help displace the leaving group.

Acid halide formation

Carboxylic acids react with phosphorous trichloride (PCl$_3$), phosphorous pentachloride (PCl$_5$), thionyl chloride (SOCl$_2$), and phosphorous tribromide (PBr$_3$) to form acyl halides.

Acid anhydride formation

Following is the anhydride group:

This group forms by reacting the salt of a carboxylic acid with an acyl halide.

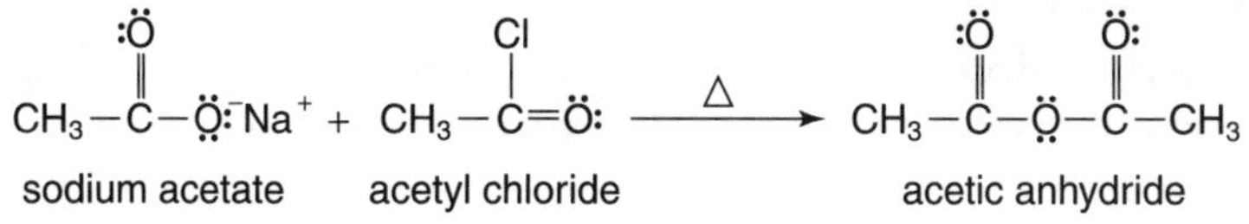

$$CH_3-\overset{\overset{\displaystyle :\overset{..}{O}}{\|}}{C}-\overset{..}{\underset{..}{O}}:Na^+ \ + \ CH_3-\overset{\overset{\displaystyle Cl}{|}}{C}=\overset{..}{O}: \ \xrightarrow{\ \triangle\ } \ CH_3-\overset{\overset{\displaystyle :\overset{..}{O}}{\|}}{C}-\overset{..}{\underset{..}{O}}-\overset{\overset{\displaystyle \overset{..}{O}:}{\|}}{C}-CH_3$$

sodium acetate   acetyl chloride   acetic anhydride

Decarboxylation reaction

Decarboxylation is the loss of the acid functional group as carbon dioxide from a carboxylic acid. The reaction product is usually a halocompound or an aliphatic or aromatic hydrocarbon.

The following illustration shows the sodalime method:

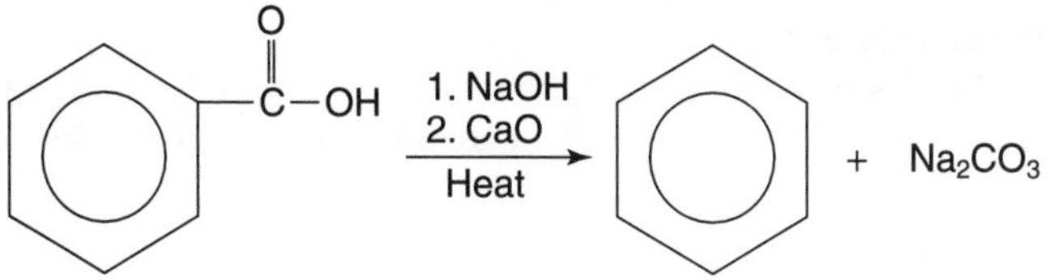

Alipathic and aromatic acids can be decarboxylated using simple copper salts.

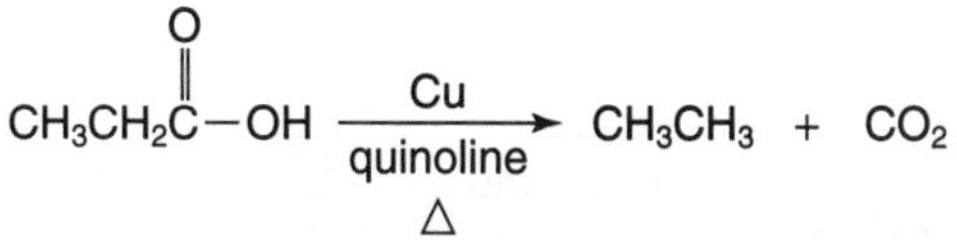

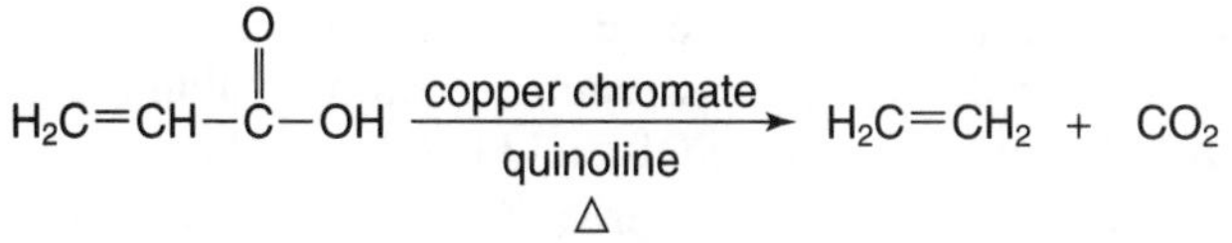

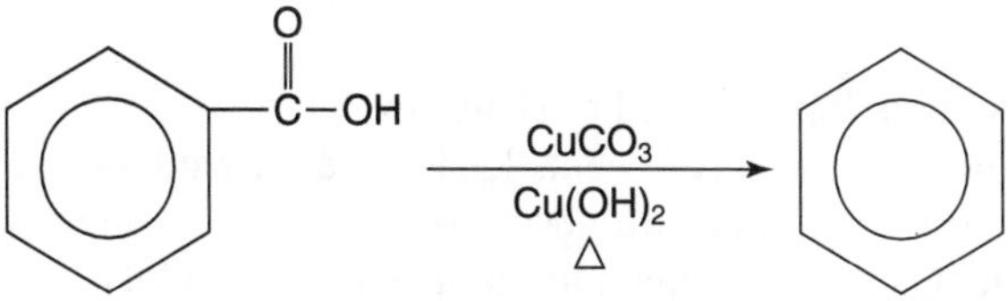

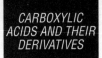

Hunsdiecker reaction

In a **Hunsdiecker reaction**, the silver salt of an aromatic carboxylic acid is converted by bromine treatment to an acyl halide.

$$\text{C}_6\text{H}_5\text{-C(=O)-O}^-\text{Ag}^+ \xrightarrow[\text{CCl}_4]{\text{Br}_2} \text{C}_6\text{H}_5\text{-Br} + \text{CO}_2 + \text{AgBr}$$

Kolbe electrolysis

In **Kolbe electrolysis,** electrochemical oxidation occurs in aqueous sodium hydroxide solution, leading to the formation of a hydrocarbon.

$$CH_3CH_2\overset{O}{\overset{\|}{C}}-O^-Na^+ \xrightarrow[\text{oxidation}]{\text{electrochemical}} CH_3CH_2CH_2CH_3 + CO_2$$

Reduction of Carboxylic Acids and Acid Derivatives

Carboxylic acids, acid halides, esters, and amides are easily reduced by strong reducing agents, such as lithium aluminum hydride (LiAlH$_4$). The carboxylic acids, acid halides, and esters are reduced to alcohols, while the amide derivative is reduced to an amine.

Reductions of carboxylic acid derivatives

Most reductions of carboxylic acids lead to the formation of primary alcohols. These reductions are normally carried out using a strong reducing agent, such as lithium aluminum hydride (LiAlH$_4$).

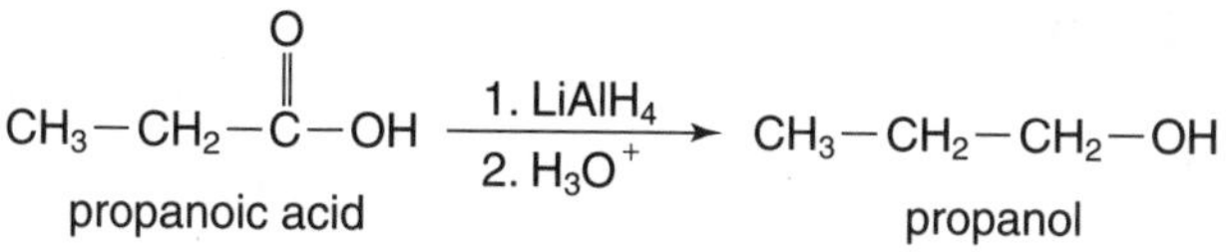

$$CH_3-CH_2-\overset{\overset{\displaystyle O}{\|}}{C}-OH \quad \xrightarrow[\text{2. } H_3O^+]{\text{1. LiAlH}_4} \quad CH_3-CH_2-CH_2-OH$$

propanoic acid   propanol

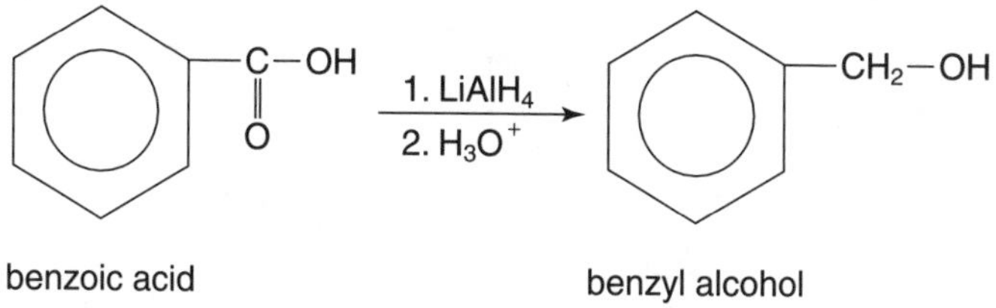

benzoic acid   benzyl alcohol

You can also use diborane (B_2H_6) to reduce carboxylic acids to alcohols.

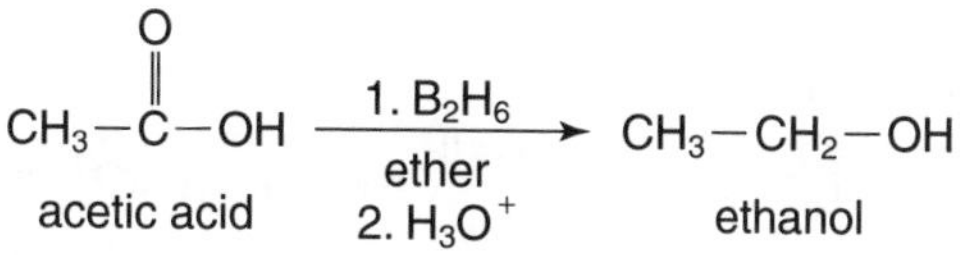

$$CH_3-\overset{\overset{\displaystyle O}{\|}}{C}-OH \quad \xrightarrow[\substack{\text{ether} \\ \text{2. } H_3O^+}]{\text{1. } B_2H_6} \quad CH_3-CH_2-OH$$

acetic acid   ethanol

Reduction of esters

Esters are normally reduced by reaction with lithium aluminum hydride.

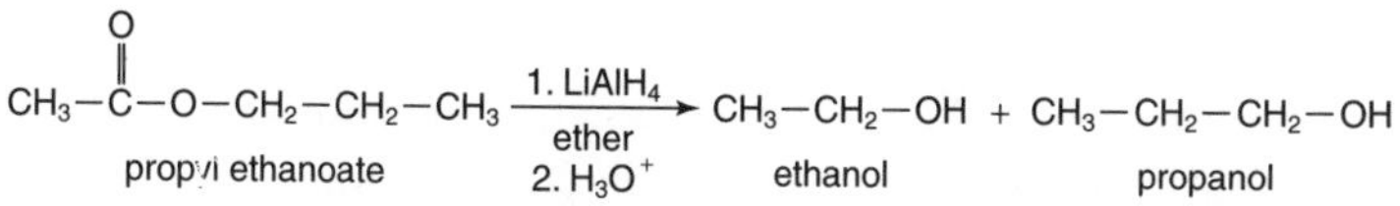

$$CH_3-\overset{\overset{\displaystyle O}{\|}}{C}-O-CH_2-CH_2-CH_3 \quad \xrightarrow[\substack{\text{ether} \\ \text{2. } H_3O^+}]{\text{1. LiAlH}_4} \quad CH_3-CH_2-OH \ + \ CH_3-CH_2-CH_2-OH$$

propyl ethanoate   ethanol   propanol

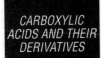

Reduction of acid halides

Acid halides are reduced by lithium aluminum hydride to primary alcohols.

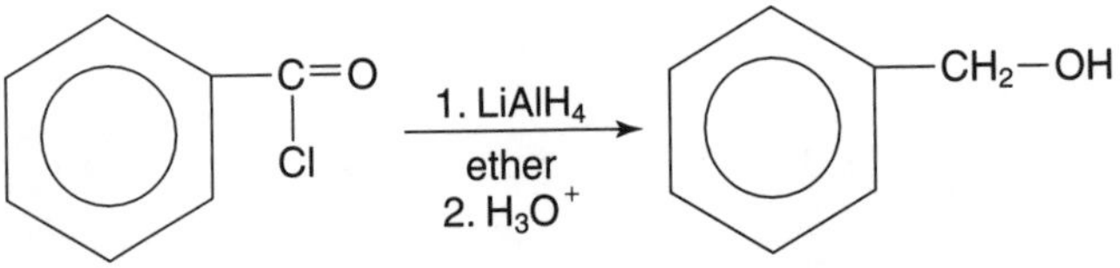

$$
\text{benzoyl chloride} \quad \xrightarrow[\substack{\text{ether} \\ 2.\ H_3O^+}]{1.\ LiAlH_4} \quad \text{benzyl alcohol}
$$

benzoyl chloride $\qquad$ benzyl alcohol

Reduction of amides

Like other carboxylic acid derivatives, amides can be reduced by lithium aluminum hydride. The product of this reduction is an amine.

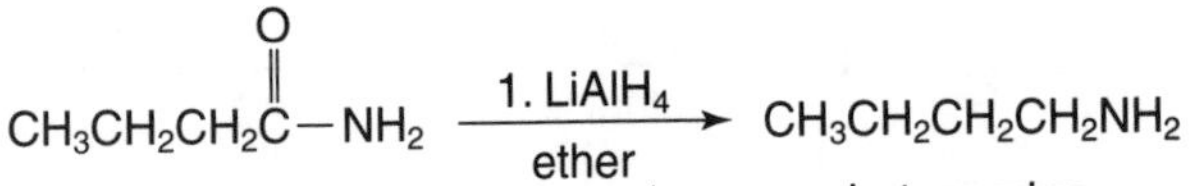

$$
\underset{\text{propanamide}}{CH_3CH_2CH_2\overset{\overset{\displaystyle O}{\|}}{C}-NH_2} \quad \xrightarrow[\substack{\text{ether} \\ 2.\ H_3O^+}]{1.\ LiAlH_4} \quad \underset{\text{butanamine}}{CH_3CH_2CH_2CH_2NH_2}
$$

Reactions of carboxylic acid derivatives

Carboxylic acid derivatives are very reactive. The following sections detail how the various carboxylic acid derivatives can be converted one into another.

Reactions of acid halides (acyl halides). Acyl halides are very reactive and easily converted to esters, anhydrides, amides, N-substituted amides, and carboxylic acids. In the following reactions, X represents any halide.

An acid halide can be converted to an ester by an acid catalyzed reaction with an alcohol.

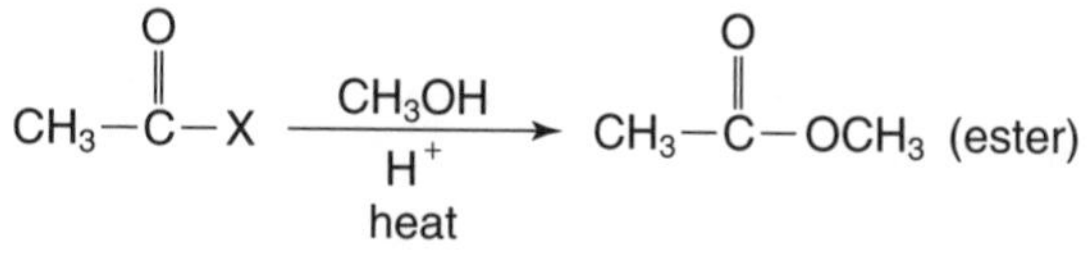

$$CH_3-\underset{\underset{O}{\|}}{C}-X \xrightarrow[\text{heat}]{\overset{CH_3OH}{\underset{H^+}{}}} CH_3-\underset{\underset{O}{\|}}{C}-OCH_3 \text{ (ester)}$$

An anhydride may be produced by reacting an acid halide with the sodium salt of a carboxylic acid.

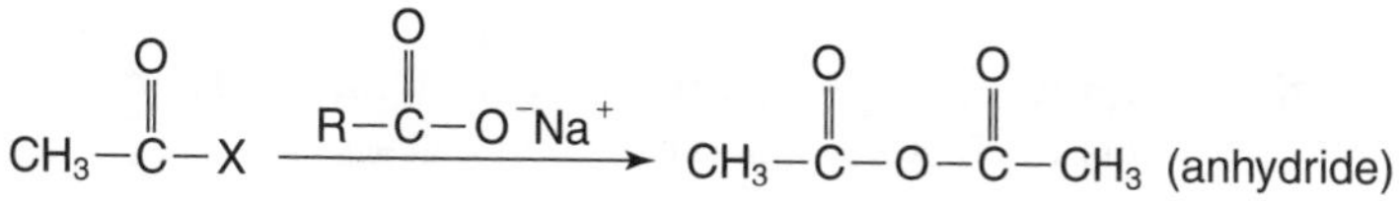

$$CH_3-\underset{\underset{O}{\|}}{C}-X \xrightarrow{R-\underset{\overset{\|}{O}}{C}-O^-Na^+} CH_3-\underset{\underset{O}{\|}}{C}-O-\underset{\underset{O}{\|}}{C}-CH_3 \text{ (anhydride)}$$

Reacting ammonia with an acid halide produces an amide.

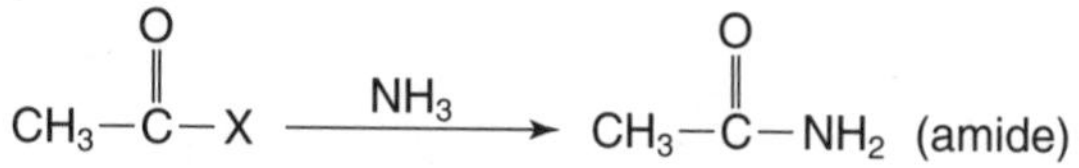

$$CH_3-\underset{\underset{O}{\|}}{C}-X \xrightarrow{NH_3} CH_3-\underset{\underset{O}{\|}}{C}-NH_2 \text{ (amide)}$$

Reacting a primary amine with an acid halide creates an N-substituted amide.

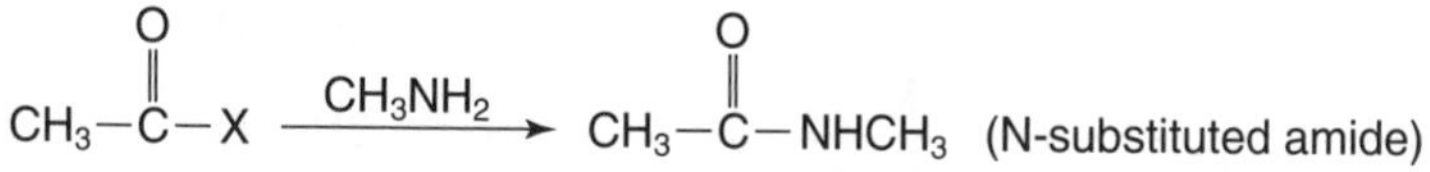

$$CH_3-\underset{\underset{O}{\|}}{C}-X \xrightarrow{CH_3NH_2} CH_3-\underset{\underset{O}{\|}}{C}-NHCH_3 \text{ (N-substituted amide)}$$

Similarly, reacting a secondary amine with an acid halide produces an N,N-disubstituted amide.

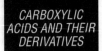
$$CH_3-\overset{O}{\overset{\|}{C}}-X \xrightarrow{(CH_3)_2NH} CH_3-\overset{O}{\overset{\|}{C}}-N(CH_3)_2 \text{ (N,N-disubstituted amide)}$$

Finally, hydrolysis of an acid halide with dilute aqueous acid produces a carboxylic acid.

$$CH_3-\overset{O}{\overset{\|}{C}}-X \xrightarrow[\substack{H^+ \\ heat}]{H_2O} CH_3-\overset{O}{\overset{\|}{C}}-OH \text{ (acid)}$$

Reaction of anhydrides. Anhydrides react rapidly to form esters, amides, N-substituted amides, and carboxylic acids.

Reaction of an alcohol with an anhydride creates an ester and a carboxylic acid.

$$CH_3-\overset{O}{\overset{\|}{C}}-\ddot{O}-\overset{O}{\overset{\|}{C}}-CH_3 \xrightarrow{CH_3OH} CH_3-\overset{O}{\overset{\|}{C}}-OCH_3 + CH_3-\overset{O}{\overset{\|}{C}}-OH$$

Reacting an anhydride with ammonia produces an amide and a carboxylic acid salt.

$$CH_3-\overset{O}{\overset{\|}{C}}-O-\overset{O}{\overset{\|}{C}}-CH_3 \xrightarrow[heat]{\ddot{N}H_3} CH_3-\overset{O}{\overset{\|}{C}}-NH_2 + CH_3-\overset{O}{\overset{\|}{C}}-O^-NH_4^+$$

A primary amine reacts with an anhydride to give an N-substituted amide.

$$CH_3-\overset{\overset{\displaystyle O}{\|}}{C}-O-\overset{\overset{\displaystyle O}{\|}}{C}-CH_3 \xrightarrow{CH_3\ddot{N}H_3} CH_3-\overset{\overset{\displaystyle O}{\|}}{C}-NHCH_3 \ + \ CH_3-\overset{\overset{\displaystyle O}{\|}}{C}-O^-NH_4{}^+$$

Similarly, a secondary N-substituted amine reacts with an anhydride to produce an N,N-disubstituted amide plus a carboxylic acid salt.

$$CH_3-\overset{\overset{\displaystyle O}{\|}}{C}-O-\overset{\overset{\displaystyle O}{\|}}{C}-CH_3 \xrightarrow{(CH_3)_2NH} CH_3-\overset{\overset{\displaystyle O}{\|}}{C}-N(CH_3)_2 \ + \ CH_3-\overset{\overset{\displaystyle O}{\|}}{C}-O^-NH_4{}^+$$

Finally, reacting an N,N-disubstituted amide anhydride with dilute aqueous acid produces a carboxylic acid.

$$CH_3-\overset{\overset{\displaystyle O}{\|}}{C}-O-\overset{\overset{\displaystyle O}{\|}}{C}-CH_3 \xrightarrow[H^+]{H_2O} CH_3-\overset{\overset{\displaystyle O}{\|}}{C}-OH$$

Reactivity of carboxylic acid derivatives

The conversion of one type of derivative into another occurs via nucleophilic acyl substitution reactions. In these types of reactions, any factor that makes the carbonyl group more easily attacked by a nucleophile favors the reaction. The two most important factors are steric hindrance and electronic factors.

Sterically unhindered, accessible carbonyl groups react more rapidly with nucleophiles than do hindered carbonyl groups. Electronically, groups which help polarize the carbonyl group make the compound more reactive. Thus acid chlorides would be more reactive than esters, because the chlorine atom is much more electronegative than an alkoxide ion.

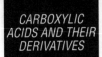
Based on the above factors, the order of reactivity of carboxylic acid derivatives is

$$
\underset{\text{acid halide}}{R-\overset{\displaystyle O}{\overset{\|}{C}}-X} \quad > \quad \underset{\text{anhydride}}{R-\overset{\displaystyle O}{\overset{\|}{C}}-\overset{..}{O}-\overset{\displaystyle O}{\overset{\|}{C}}-R} \quad > \quad \underset{\text{ester}}{R-\overset{\displaystyle O}{\overset{\|}{C}}-\overset{..}{O}R} \quad > \quad \underset{\text{amide}}{R-\overset{\displaystyle O}{\overset{\|}{C}}-NH_2}
$$

The more reactive acid derivative can be easily converted into a less reactive derivative. However, the opposite cannot occur. Thus, less reactive derivatives cannot be converted into their more reactive cousins.

Introduction

Amines are aliphatic and aromatic derivatives of ammonia. Amines, like ammonia, are weak bases ($K_b = 10^4$ to 10^6). This basicity is due to the unshared electron pair on the nitrogen atom.

Classification and nomenclature of amines

Amines are classified as primary, secondary, or tertiary based upon the number of carbon-containing groups that are attached to the nitrogen atom. Those amine compounds that have only one group attached to the nitrogen atom are primary, while those with two or three groups attached to the nitrogen atom are secondary and tertiary, respectively.

$CH_3\ddot{N}H_2$

1°
methanamine
(methylamine)

$CH_3 — \underset{\underset{H}{|}}{N} — CH_3$

2°
N-methylmethanamine
(dimethylamine)

$CH_3 — \underset{\underset{CH_3}{|}}{N} — CH_3$

3°
N, N-dimethylmethanamine
(trimethylamine)

In the common system, you name amines by naming the group or groups attached to the nitrogen atom and adding the word amine.

$CH_3\ddot{N}H_2$

methyl amine

$CH_3 — \underset{\underset{H}{|}}{N} —$

methyl phenyl amine

$CH_3 — \underset{\underset{CH_3}{|}}{N} — CH_3$

trimethyl amine

In the IUPAC System, apply the following rules to name amines:

1. Pick out the longest continuous chain of carbon atoms. The parent name comes from the alkane of the same number of carbons.

2. Change the -e of the alkane to "amine."

3. Locate and name any substituents, keeping in mind that the chain is numbered away from the amine group. Substituents, which are attached to the nitrogen atom instead of the carbon of the chain, are designated by a capital N.

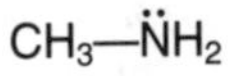

$CH_3-\ddot{N}H_2$

methanamine

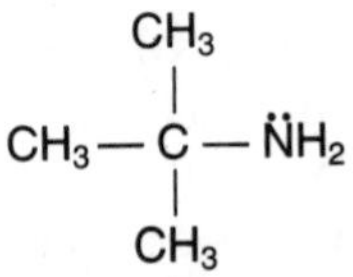

$CH_3-\overset{\displaystyle CH_3}{\underset{\displaystyle CH_3}{\overset{|}{\underset{|}{C}}}}-\ddot{N}H_2$

1, 1-dimethylethanamine

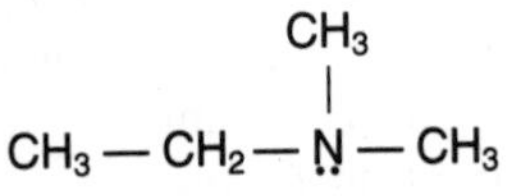

$CH_3-CH_2-\overset{\displaystyle CH_3}{\overset{|}{N}}-CH_3$

N, N-dimethylethanamine

Aromatic amines belong to specific families, which act as parent molecules. For example, an amino group (—NH₂) attached to benzene produces the parent compound aniline.

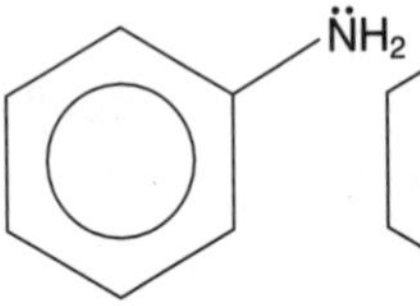

aniline
(parent)

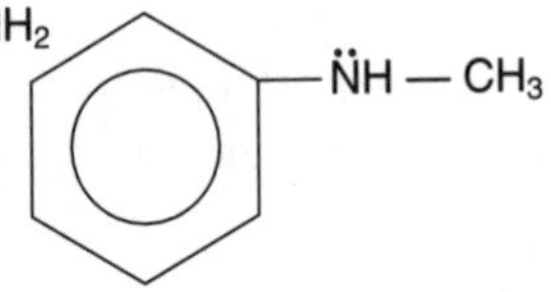

N-methylaniline
(N-substituted)

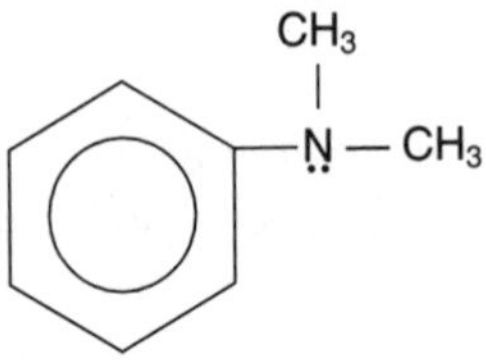

N, N-dimethylaniline
(N, N-disubstituted)

Basicity of amines

Amines are basic because they possess a pair of unshared electrons, which they can share with other atoms. These unshared electrons create an electron density around the nitrogen atom. The greater the electron density, the more basic the molecule. Groups that donate or supply electrons will increase the basicity of amines while groups that decrease the electron density around the nitrogen decrease the basicity of the molecule. For alkyl halides in the gas phase, the order of base strength is given below:

$$(CH_3)_3 N > (CH_3)_2 NH > CH_3NH_2 > NH_3$$

most		least
basic		basic

However, in aqueous solutions, the order of basicity changes.

$$(CH_3)_2 NH > CH_3NH_2 > (CH_3)_3 N > NH_3$$

most		least
basic		basic

The differences in the basicity order in the gas phase and aqueous solutions are the result of solvation effects. Amines in water solution exist as ammonium ions.

$$R - \ddot{N}H_2 + H_2O \longrightarrow RHNH_3^+ \bar{O}H$$

1° amine   1° ammonium salt

$$R - \ddot{N}HR + H_2O \longrightarrow RNH_2R^+ \bar{O}H$$

2° amine   2° ammonium salt

$$R\ddot{N}R_2 + H_2O \longrightarrow RNHR_2^+ \bar{O}H$$

3° amine   3° ammonium salt

In water, the ammonium salts of primary and secondary amines undergo solvation effects (due to hydrogen bonding) to a much greater degree than ammonium salts of tertiary amines. These solvation effects increase the electron density on the amine nitrogen to a greater degree than the inductive effect of alkyl groups.

Arylamines are weaker bases than cyclohexylamines because of resonance. Aniline, a typical arylamine, exhibits the resonance structures shown in Figure 8-1.

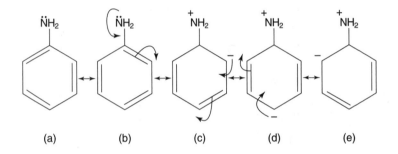

Figure 8-1

As structures b through e in Figure 8-1 show, delocalization of the unshared electron pair occurs throughout the ring, making these electrons less available for reaction. As a result of this electron delocalization, the molecule becomes less basic.

Preparation of Amines

The alkylation of ammonia, Gabriel synthesis, reduction of nitriles, reduction of amides, reduction of nitrocompounds, and reductive amination of aldehydes and ketones are methods commonly used for preparing amines.

Alkylation of ammonia

The reaction of ammonia with an alkyl halide leads to the formation of a primary amine. The primary amine that is formed can also react with the alkyl halide, which leads to a disubstituted amine that can further react to form a trisubstituted amine. Therefore, the alkylation of ammonia leads to a mixture of products.

$$\ddot{N}H_3 + CH_3Cl \longrightarrow CH_3\overset{+}{N}H_4\overset{-}{Cl} \xrightarrow{OH^-} CH_3\ddot{N}H_2$$

alkylation of ammonia

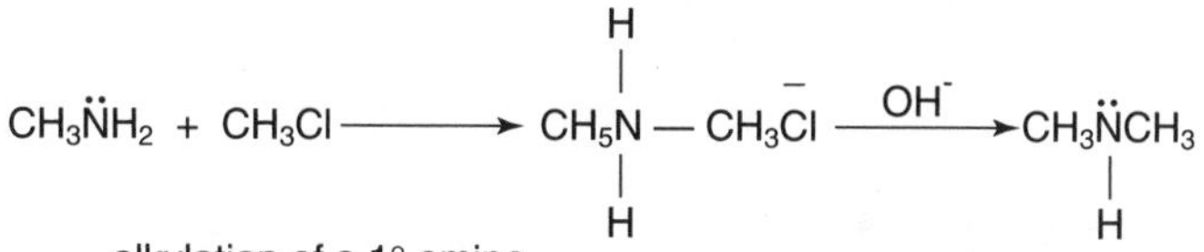

$$CH_3\ddot{N}H_2 + CH_3Cl \longrightarrow CH_5\overset{\overset{H}{|}}{\underset{\underset{H}{|}}{N}} - CH_3\overset{-}{Cl} \xrightarrow{OH^-} CH_3\underset{\underset{H}{|}}{\ddot{N}}CH_3$$

alkylation of a 1° amine

$$CH_3\ddot{N}CH_3 + CH_3Cl \longrightarrow CH_3 - \overset{\overset{CH_3}{|}}{\underset{\underset{CH_3}{|}}{\overset{+}{N}}} - CH_3\overset{-}{Cl} \xrightarrow{OH^-} CH_3 - \underset{\underset{CH_3}{|}}{\ddot{N}} - CH_3$$

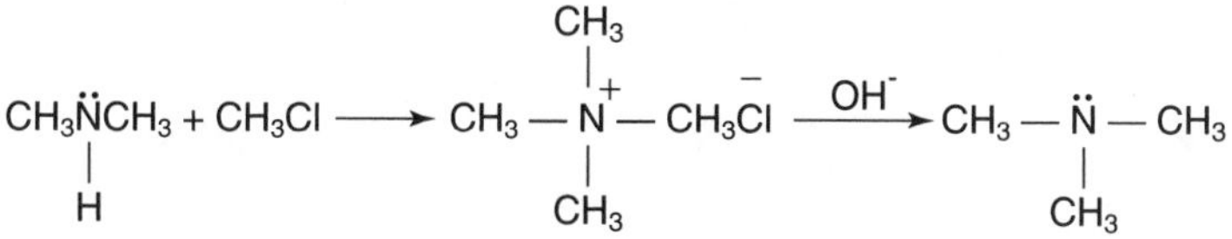

alkylation of a 2° amine

Reduction of alkylazides

You can best prepare a primary amine from its alkylazide by reduction or by the Gabriel synthesis.

$$CH_3CH_2Cl \xrightarrow[H_2O]{NaN_3} CH_3CH_2N_3 \xrightarrow[Alcohol]{Na} CH_3CH_2NH_2$$

Ethyl Azide

In the **Gabriel synthesis,** potassium phthalimide is reacted with an alkyl halide to produce an N-alkyl phthalimide. This N-alkyl phthalimide can be hydrolyzed by aqueous acids or bases into the primary amine.

potassium phthalimide   N-ethyl phthalimide   sodium phthalate

Reduction of nitriles

Nitriles can be reduced by lithium aluminum hydride ($LiAlH_4$) to primary amines.

Reduction of amides

Amides yield primary amines on reduction by lithium aluminum hydride, while N-substituted and N, N-disubstituted amides produce secondary and tertiary amines, respectively.

$$CH_3 - CH_2 - \overset{\overset{\displaystyle O}{\|}}{C} - NH_2 \xrightarrow[\text{2. } H_3O^+]{\text{1. LiAlH}_4 \quad \text{ether}} CH_3 - CH_2 - CH_2 - NH_2$$

propanamide   propanamine

$$CH_3 - CH_2 - \overset{\overset{\displaystyle O}{\|}}{C} - NH - CH_3 \xrightarrow[\text{2. } H_3O^+]{\text{1. LiAlH}_4 \quad \text{ether}} CH_3 - CH_2 - CH_2 - NH - CH_3$$

N-methylpropanamide   N-methylpropanamine

$$CH_3 - CH_2 - \overset{\overset{\displaystyle O}{\|}}{C} - \overset{\overset{\displaystyle CH_3}{|}}{N} - CH_3 \xrightarrow[\text{2. } H_3O^+]{\text{1. LiAlH}_4 \quad \text{ether}} CH_3 - CH_2 - CH_2 - \overset{\overset{\displaystyle CH_3}{|}}{N} - CH_3$$

N, N-dimethylpropanamide   N, N-dimethylpropanamine

Because amides are easily prepared, their reduction is a preferred method for making all classes of amines.

Reduction of nitrocompounds

Aromatic amines are normally prepared by reduction of the corresponding aromatic nitrocompound.

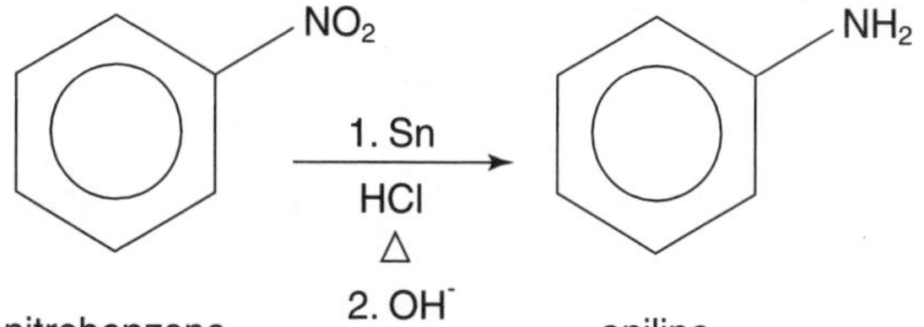

nitrobenzene   aniline

Reductive amination of aldehydes and ketones

Aldehydes or ketones can be reduced by catalytic or chemical reductions in the presence of ammonia or primary or secondary amines, producing primary, secondary, or tertiary amines.

The reaction of a ketone with ammonia, followed by catalytic reduction or reduction by sodium cyanoborohydride, produces a 1° amine.

$$CH_3 - \overset{\overset{\displaystyle O}{\|}}{C} - CH_3 \xrightarrow[\substack{\text{or} \\ \text{NaBH}_3\text{CN}}]{\substack{1.\ NH_3 \\ 2.\ H_2/Ni}} CH_3 - \overset{\overset{\displaystyle CH_3}{|}}{\underset{\underset{\displaystyle H}{|}}{C}} - NH_2$$

N-substituted amines are produced by reaction of ketones with primary amines, followed by reduction.

$$CH_3 - \overset{\overset{\displaystyle O}{\|}}{C} - CH_3 \xrightarrow[\substack{\text{or} \\ \text{NaBH}_3\text{CN}}]{\substack{1.\ CH_3NH_2 \\ 2.\ H_2/Ni}} CH_3 - \overset{\overset{\displaystyle CH_3}{|}}{\underset{\underset{\displaystyle H}{|}}{C}} - NHCH_3$$

N,N-disubstituted amines can be produced by reaction of 2° amines with ketones followed by reduction.

$$CH_3 - \overset{\overset{\displaystyle O}{\|}}{C} - CH_3 \xrightarrow[\substack{\text{or} \\ \text{NaBH}_3\text{CN}}]{\substack{1.\ (CH_3)_2NH \\ 2.\ H_2/Ni}} CH_3 - \overset{\overset{\displaystyle CH_3}{|}}{\underset{\underset{\displaystyle H}{|}}{C}} - N(CH_3)_2$$

Reactions of Amines

Due to the unshared electron pair, amines can act as both bases and nucleophiles.

Reaction with acids

When reacted with acids, amines donate electrons to form ammonium salts.

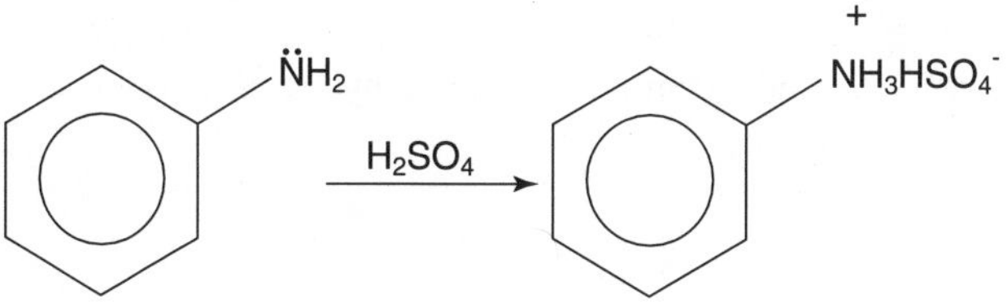

aniline   anilinium acid sulfate

Reaction with acid halides

Acid halides react with amines to form substituted amides.

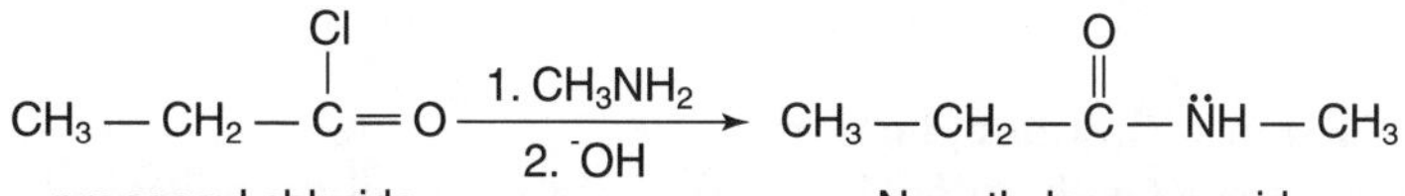

propanoyl chloride   N-methylpropanamide

Reaction with aldehydes and ketones

Aldehydes and ketones react with primary amines to give a reaction product (a carbinolamine) that dehydrates to yield aldimines and ketimines (Schiff bases).

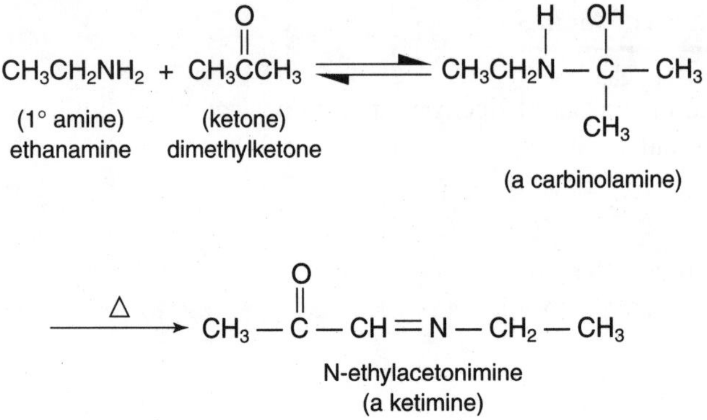

$$CH_3CH_2NH_2 + CH_3\overset{\overset{\displaystyle O}{\|}}{C}CH_3 \rightleftharpoons CH_3CH_2\overset{\overset{\displaystyle H}{|}}{N} - \overset{\overset{\displaystyle OH}{|}}{\underset{\underset{\displaystyle CH_3}{|}}{C}} - CH_3$$

(1° amine)   (ketone)
ethanamine   dimethylketone

(a carbinolamine)

$$\xrightarrow{\Delta} \quad CH_3 - \overset{\overset{\displaystyle O}{\|}}{C} - CH = N - CH_2 - CH_3$$

N-ethylacetonimine
(a ketimine)

If you react secondary amines with aldehydes or ketones, enamines form.

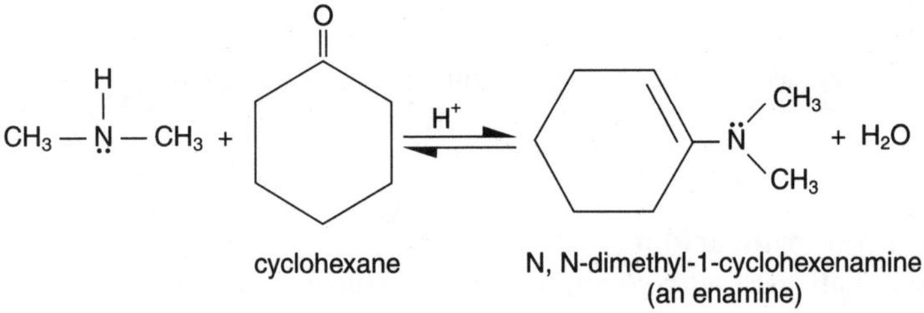

$$CH_3 - \overset{\overset{\displaystyle H}{|}}{\underset{\cdot\cdot}{N}} - CH_3 \; + \quad \xrightleftharpoons{H^+} \quad + \; H_2O$$

cyclohexane   N, N-dimethyl-1-cyclohexenamine
(an enamine)

Reaction with sulfonyl chlorides

Amines react with sulfonyl chlorides to produce sulfonamides. A typical example is the reaction of benzene sulfonyl chloride with aniline.

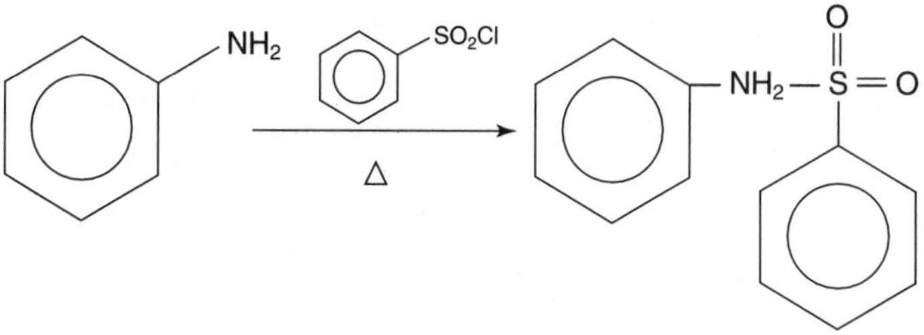

N-phenylbenzenesulfonamide

The Hinsberg test

The Hinsberg test, which can distinguish primary, secondary, and tertiary amines, is based upon sulfonamide formation. In the Hinsberg test, an amine is reacted with benzene sulfonyl chloride. If a product forms, the amine is either a primary or secondary amine, because tertiary amines do not form stable sulfonamides. If the sulfonamide that forms dissolves in aqueous sodium hydroxide solution, it is a primary amine. If the sulfonamide is insoluble in aqueous sodium hydroxide, it is a secondary amine. The sulfonamide of a primary amine is soluble in an aqueous base because it still possesses an acidic hydrogen on the nitrogen, which can be lost to form a sodium salt.

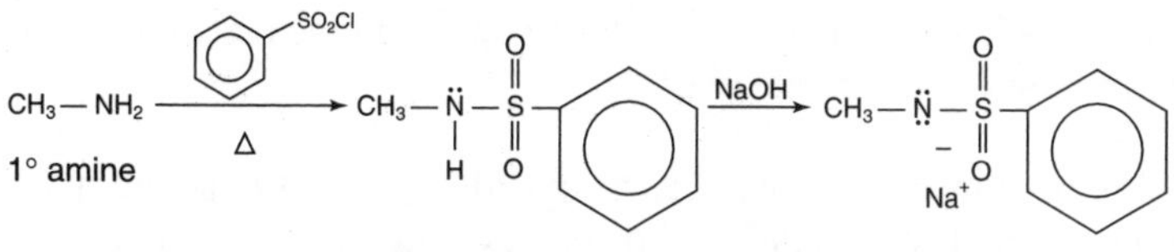

water insoluble   water soluble

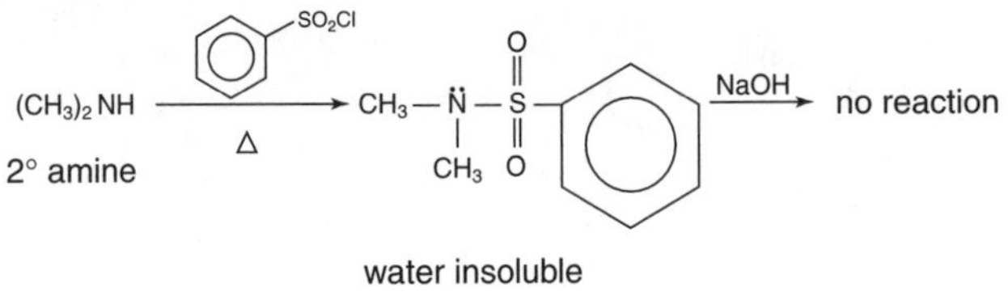

water insoluble

Oxidation

Although you can oxidize all amines, only tertiary amines give easily isolated products. The oxidation of a tertiary amine leads to the formation of an amine oxide.

$$(CH_3)_3N \xrightarrow[\substack{\text{or} \\ \text{peroxyacetic} \\ \text{acid}}]{H_2O_2} (CH_3)_3N^+ - O^-$$

Arylamines tend to be easily oxidized, with oxidation occurring on the amine group as well as in the ring.

Reaction with nitrous acid

Nitrous acid is unstable and must be prepared in the reaction solution by mixing sodium nitrite with acid.

Primary amines react with nitrous acid to yield a diazonium salt, which is highly unstable and degrades into a carbocation that is capable of reaction with any nucleophile in solution. Therefore, reacting primary amines with nitrous acid leads to a mixture of alcohol, alkenes, and alkyl halides.

$$CH_3CH_2NH_2 \xrightarrow[\text{HCl}]{\text{NaNO}_2} CH_3CH_2 \overset{+}{-} N \equiv NCl^-$$

ethanamine   0°   ethyl diazonium chloride
(ethylamine)   (unstable)

$$CH_3CH_2\overset{+}{N} \equiv NCl^- \longrightarrow \left[CH_3CH_2^+\right] + N_2 \longrightarrow$$

$$\xrightarrow{\text{HOH}} CH_3CH_2OH$$
$$\xrightarrow{\text{Cl}^-} CH_3CH_2Cl$$
$$\longrightarrow CH_2 = CH_2 + H^+$$

Primary aromatic amines form stable diazonium salts at zero degrees.

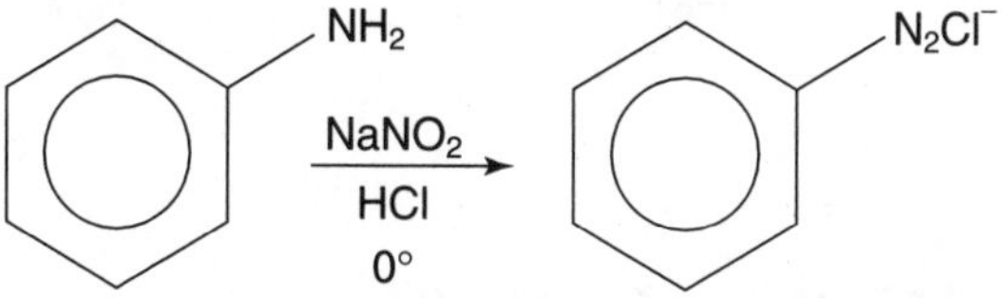

aniline   benzene diazonium chloride

Secondary aliphatic and aromatic amines form nitrosoamine with nitrous acid.

$$CH_3 - NH - CH_3 \xrightarrow[\text{HCl}]{\text{NaNO}_2} (CH_3)_2 - N - N = O$$

dimethylamine   N,nitrosodimethylamine

N-methylaniline   N-nitroso N-methylaniline

Tertiary amines react with nitrous acid to form N-nitrosoammonium compounds.

Reactions of aromatic diazonium salts

Diazonium salts of aromatic amines are very useful as intermediates to other compounds. Because aromatic diazonium salts are only stable at very low temperatures (zero degrees and below), warming these salts initiates decomposition into highly reactive cations. These cations can react with any anion present in solution to form a variety of compounds. Figure 8-2 illustrates the diversity of the reactions.

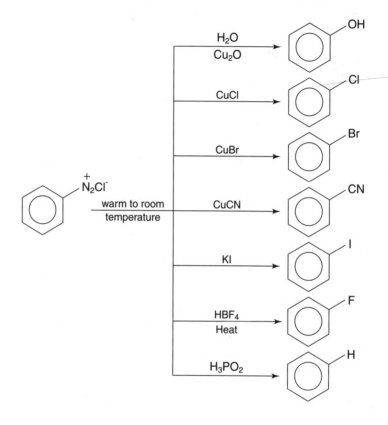

Figure 8-2

Introduction

The interpretation of data generated from instrumentation allows organic chemists to assign possible structures to new molecules or to identify existing materials. The data provided by instruments is usually in the form of graphs called **spectra.** To interpret spectra, a chemist must have some knowledge of the phenomena occurring within a molecule when varying amounts of energy are added to the molecule. The source of the added energy can come from interacting with high-energy electrons (mass spectroscopy), radio waves (nuclear magnetic spectroscopy), light energy (ultraviolet and visible spectroscopy), and heat energy (infrared spectroscopy).

Mass Spectra

In a mass spectrometer, a solid or liquid is heated under reduced pressure to convert it into a gas. The molecules of the gas are then exposed to high-energy electrons. Collisions between the gaseous molecules and the electrons convert some of the molecules into positively charged ions. These ions are passed through magnetic and electrical fields that separate them into a spectrum based on their mass-to-charge ratio (m/z). After separation, the ions arrive at a detector that is calibrated to receive only ions of a plus one charge. The spectrum that results gives the molecular weights of the ions.

Integral molecular weight

The most important information obtainable from a mass spectrum is the **integral molecular weight** of the compound. The integral molecular weight is the sum of the light isotopes of the atoms that make up the molecule. The integral masses of atoms used to calculate integral molecular weights are listed below in Table 9-1.

Table 9-1: Integral Atomic Weights

Atom	Atomic Weight
Hydrogen	1
Carbon	12
Nitrogen	14
Oxygen	16
Fluorine	19
Phosphorous	31
Sulfur	32
Chloride	35
Bromine	79
Iodine	127

The molecular weight of the compound appears as the molecular ion peak on the spectrum. Ordinarily, the molecular ion peak is followed in the spectrum by two peaks; one of m/z one mass unit greater than the molecular ion and a second peak of m/z two mass units greater than the molecular ion. These peaks are called the **molecular ion + 1** and the **molecular ion + 2** peaks. The intensity of the molecular ion, molecular ion + 1, and molecular ion + 2 peaks exist in an approximate ratio of 100:10:1 for many hydrocarbons or molecules possessing one or more hydrocarbon chains.

The appearance of this type of peak arrangement at the high end of the m/z scale on the mass spectrum is characteristic of the molecular ion. The two higher mass peaks are due to the presence of isotopes of the compound's atoms. The ratio found in the peak pattern is the result of natural isotope abundance of the atoms in the molecule. In a number of spectra, no molecular ion can be found, because the molecular ion is usually very unstable and decomposes into fragment ions before it can reach the detector.

Fragment and rearrangement ions

In addition to forming molecular ions, organic molecules decompose into fragment ions in a mass spectrometer. As a result, a host of ions form that have a m/z less than that of the molecular ion. These ions give rise to the characteristic mass spectrum of the molecule. The fragmentation pattern of a molecule is characteristic of the molecule, and an unknown compound may be identified by comparison with a catalog of standard spectra.

Fragment ions are formed through both simple cleavage of larger ions and rearrangements of the molecular ion. Simple cleavage of the molecular ion tends to form stable cations. Thus, simple cleavage generally occurs next to highly branched carbons, at allylic and benzylic bonds, and at carbons adjacent to oxygen and nitrogen atoms.

Fragmentation can also occur by the expulsion of a neutral molecule during the rearrangement of the molecular ion. Many rearrangements involve the loss of small molecules, such as water, carbon monoxide, carbon dioxide, and ammonia.

Nuclear Magnetic Resonance (NMR) Spectra

Nuclei of atoms with an odd number of protons or neutrons have permanent magnetic moments and quantized nuclear spin states. This means that these types of atoms behave as though they are small magnets spinning on an axis. Placing these types of atoms in a very strong magnetic field separates them into two groups: those that align with the applied field—the field created by the electromagnet of the instrument—and those that align against the applied field.

Aligning against the applied field takes more energy than aligning with the applied field. When the sample is irradiated with radio waves, energy is quantitatively absorbed by the odd-numbered nuclei, and those aligned with the field will *flip* to align against the field. Depending upon the environment in which the proton is located, slightly more or less energy is necessary to create the flip. Thus, radio waves of varying frequencies are needed.

Deshielded and shielded protons

In practice, it is easier to fix the radio wave frequency and vary the applied magnetic field than it is to vary the radio wave frequency. The magnetic field "felt" by a hydrogen atom is composed of both applied and induced fields. The **induced field** is a field created by the electrons in the bond to the hydrogen and the electrons in nearby π bonds. When the two fields reinforce each other, a smaller applied field is required to flip the proton. In this situation, a proton is said to be **deshielded.** When the applied and induced fields oppose each other, a stronger field must be applied to flip the proton. In this state, the proton is **shielded.**

The following generalizations apply to shielding and deshielding of the protons in a molecule:

- Electronegative atoms such as nitrogen, oxygen, and halogens deshield hydrogens. The extent of deshielding is proportional to the electronegativity of the hetero atom and its proximity to the hydrogen.

- Electrons on an aromatic ring, double bonded atoms, and triple bonded atoms deshield attached hydrogens.

- A carbonyl group deshields hydrogens on adjacent chains.

- Benzylic and allylic hydrogens are deshielded.

- Electropositive atoms, such as silicon, shield hydrogens.

■ Hydrogens attached to a cyclopropane ring and those situated in the π cloud of an aromatic system are strongly shielded.

Chemical shifts

Changes in energy needed to flip protons are called **chemical shifts.** The location of chemical shifts (peaks) on a NMR spectrum are measured from a reference point that the hydrogens in a standard reference compound—$(CH_3)_4Si$ or tetramethylsilane (TMS)—produce. The amount of energy necessary to flip protons in TMS is assigned the arbitrary value of zero δ. Chemical shifts are measured in parts per million magnetic field strength difference (δ-scale), relative to TMS.

Deshielded protons absorb downfield on the NMR spectrum (at a lower magnetic field strength than shielded protons).

Mapping nonequivalent hydrogens

Every nonequivalent hydrogen has a unique and characteristic chemical shift that gives rise to a distinct peak or group of peaks. For example, in the propane molecule, two types of nonequivalent hydrogens exist. The first type is methyl hydrogens and the second type is methylene hydrogens. In the following diagram, methyl hydrogens are designated H_a while methylene hydrogens are designated H_b.

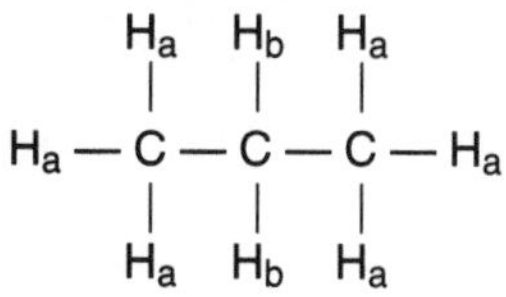

$$
\begin{array}{ccccccc}
 & H_a & & H_b & & H_a & \\
 & | & & | & & | & \\
H_a & - C & - C & - C & - H_a \\
 & | & & | & & | & \\
 & H_a & & H_b & & H_a &
\end{array}
$$

In the propene molecule, four types of nonequivalent hydrogens are designated a through d.

The H$_c$ and H$_d$ differ because H$_c$ is cis to the H$_b$ hydrogens while H$_d$ is trans.

For the benzene ring system, all hydrogens are equivalent.

Monosubstituted benzenes, however, have nonequivalent hydrogens.

This nonequivalence is due to changing environments as the hydrogens move further away from the electronegative bromine.

Peak areas

The area under a peak is directly proportional to the number of equivalent hydrogens giving rise to the signal.

Peak splitting: Spin-spin coupling

Most chemical shifts aren't single peaks but rather groups or clusters of peaks. These groups and clusters gather because of spin-spin coupling, which results from the magnetic fields of hydrogen atoms on adjacent carbon atoms reinforcing or opposing the applied magnetic field on an individual proton. In the molecule

$$X \overset{\overset{\displaystyle X}{|}}{\underset{\underset{\displaystyle H_a}{|}}{C}} \overset{\overset{\displaystyle X}{|}}{\underset{\underset{\displaystyle H_b}{|}}{C}} H_b$$

the chemical shift for the H_a atom is split into three peaks (a triplet), while the chemical shift for the H_b atoms is split into two peaks (a doublet).

The general rule for splitting is that the number of peaks created from a chemical shift is calculated as n + 1, where *n* equals the number of equivalent hydrogen atoms on the adjacent carbon atom(s) that cause the splitting. Applying this rule to the previous compound shows that the carbon adjacent to the carbon bearing the H_a hydrogen has two equivalent (H_b) hydrogens attached to it. Thus, the H_a hydrogen's chemical shift will be split into 2 + 1, or 3, peaks. The chemical shift for the H_b hydrogen atoms will be split by the single H_a hydrogen on the adjacent carbon into 1 + 1, or 2, peaks. Because the doublet represents the two H$_b$ protons and the triplet represents the single H$_a$ proton, the areas under the peaks are in a ratio of 2:1 (doublet : triplet ratio).

Coupling constants

The center-line spacing between peaks in a cluster—the space from the middle of one peak in a set to the middle of a second peak in that set–caused by spin-spin coupling is always constant. This constant value is called the **coupling constant** (J) and is expressed in hertz. The J value depends upon the structural relationship among the coupled hydrogens and is often used to help create a possible structural formula. For example, look at the following isomeric structures of the C$_2$H$_2$BrCl (bromochloroethene) compound. In any ethylene or any pair of geometric isomers, the J value will always be larger in trans arrangements than in cis arrangements. In addition, the J values will vary in a regular manner with respect to the electronegativity of the substituents.

smaller J value   larger J value

Ultraviolet and Visible Spectra

When electromagnetic radiation in the ultraviolet (UV) or visible region of the spectrum is absorbed by a molecule, π or non-bonding (n) electrons are promoted into antibonding orbitals. Because the main electron transitions are π $\longrightarrow$ π* transitions, the absorption of energy in the UV or visible range (200–700 nm) usually indicates the existence of π bond(s) and an unsaturated compound. Conversely, molecules that don't absorb in the UV or visible region don't contain an unsaturated system. The strength of the absorption, measured by its extinction coefficient, ε, is determined by the amount of conjugation in the system. The position of an absorption is determined to a large extent by the relative mobility of the π electrons.

Infrared Spectra

When a molecule absorbs energy in the infrared region (1–300 μm), the σ bonds of the molecule begin to vibrate. For simple diatomic molecules, such as H_2 or HCl, the only possible vibration is a movement of the two atoms away from and back to each other. This mode is referred to as a "bond stretch." Triatomic molecules such as CO_2 have two distinct stretching modes—an asymmetrical and a symmetrical mode. In the symmetrical stretch, both oxygen atoms move away from the carbon atom at the same time. Conversely, in the asymmetrical stretch, one oxygen atom moves toward the carbon atom while the second oxygen atom moves away from the carbon atom.

O = C = O	O = C = O
← —→	← ←
Symmetrical Stretch	Asymmetrical Stretch

Molecules of three or more atoms have continuously changing bond angles. The opening and closing of the bond angles are called **bending modes**. Some common bending modes are scissoring, rocking, twisting, and wagging. Scissoring and rocking are in-plane bending while twisting and wagging are out-of-plane bending. Figure 9-1 illustrates these vibrational modes.

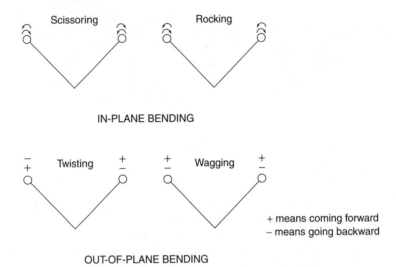

Figure 9-1

In a molecule, each bond and each group of three or more atoms absorbs infrared radiation at certain wave numbers to give quantized excited stretching and bending vibrational states. Only vibrations that cause a change in dipole moment generate an absorption peak. An observed absorption band (peak) at a specific wavelength proves the existence of a particular bond or group of bonds in the molecule. Conversely, the absence of a peak in the spectrum rules out the presence of the bond that would have produced it. The region between 1400–800 cm⁻¹ is called the **fingerprint region** of the compound. In this region, so many peaks occur that accurately identifying their origin is impossible. However, because so many peaks exist in this region, two compounds whose spectra are identical in this region must be the same.

Alkyl Halides

Addition of halogen halide to alkenes

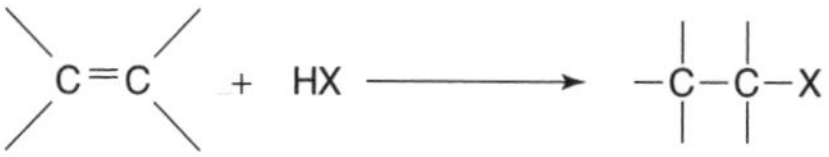

$$\text{C=C} \quad + \quad HX \quad \longrightarrow \quad -\overset{|}{\underset{|}{C}}-\overset{|}{\underset{|}{C}}-X$$

Reaction of phosphorus and sulfur halides with alcohols

$$ROH + SOX_2 \longrightarrow R-X + SO_2 + HCl$$

thionyl halide

$$ROH + PCl_3 \longrightarrow R-Cl + P(OH)_3 + HCl$$

$$ROH + PCl_5 \longrightarrow R-Cl + POCl_3 + HCl$$

Phenols

Pyrolysis of sodium benzene sulfonate

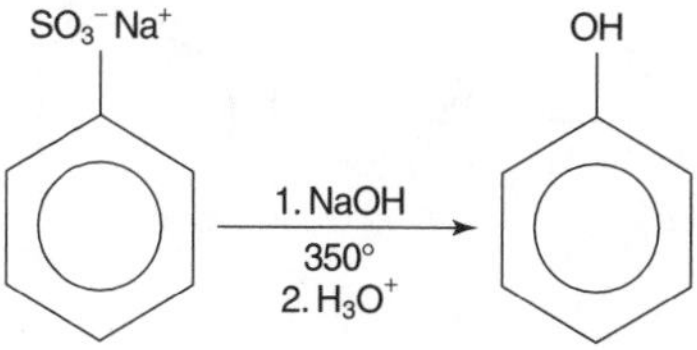

Dow process

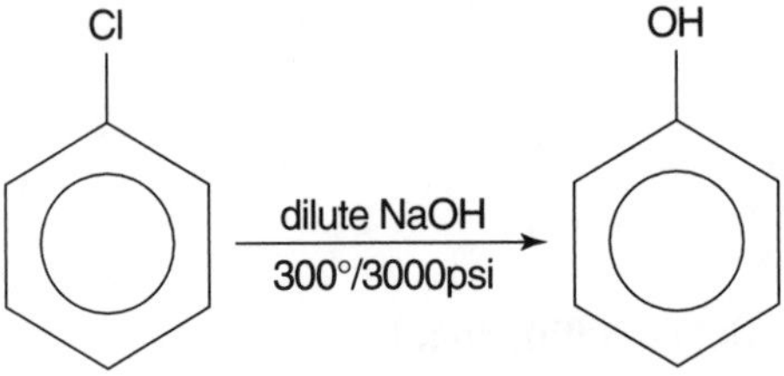

Cl → OH

dilute NaOH
300°/3000psi

Air oxidation of cumene

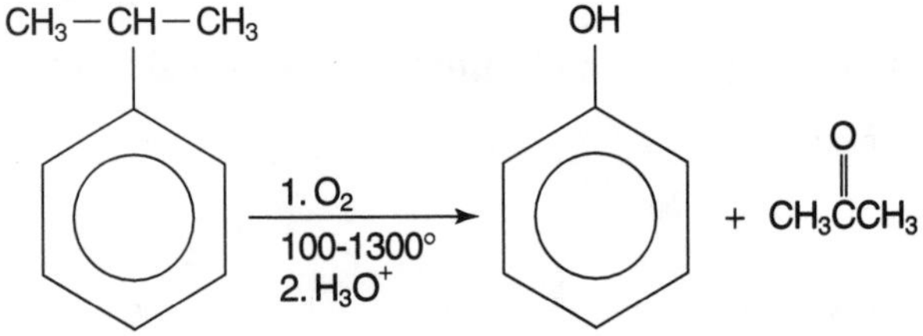

$CH_3-CH-CH_3$

1. O_2
100-1300°
2. H_3O^+

OH + $CH_3\overset{O}{\overset{\|}{C}}CH_3$

<u>Aryl Halides</u>

Halogenation of benzene

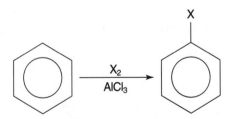

X_2
$AlCl_3$

X

Sandmeyer reaction

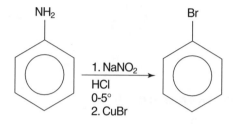

Ethers

Sulfuric acid process

$$R-OH \xrightarrow[\Delta]{H_2SO_4} R-O-R$$

Williamson synthesis

$$RO^-Na^+ \xrightarrow[\Delta]{R'X} ROR'$$

Alcohols

Hydration of alkenes

$$R-CH=CH_2 \xrightarrow[H_3O^+]{H_2O} R-\overset{\overset{\displaystyle OH}{|}}{CH}-CH_3$$

Hydroboration-oxidation

$$R-CH=CH_2 \xrightarrow[\substack{2.\ H_2O_2 \\ ^-OH}]{1.\ BH_3} R-CH_2CH_2OH$$

Reduction of aldehydes and ketones

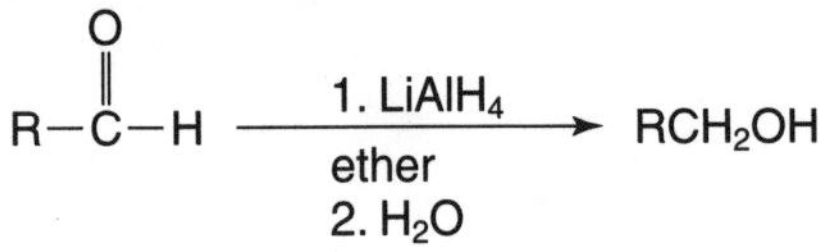

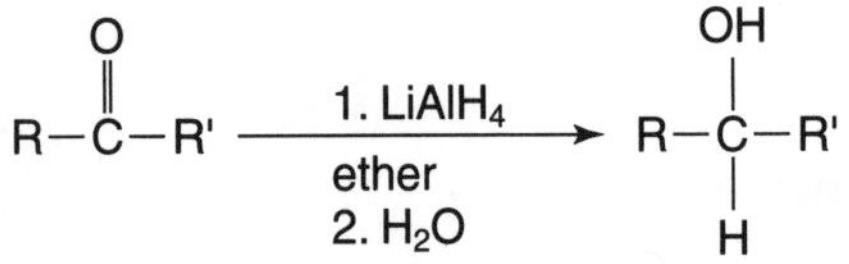

$$R-\overset{\overset{\displaystyle O}{\|}}{C}-H \xrightarrow[\substack{ether \\ 2.\ H_2O}]{1.\ LiAlH_4} RCH_2OH$$

$$R-\overset{\overset{\displaystyle O}{\|}}{C}-R' \xrightarrow[\substack{ether \\ 2.\ H_2O}]{1.\ LiAlH_4} R-\overset{\overset{\displaystyle OH}{|}}{\underset{\underset{\displaystyle H}{|}}{C}}-R'$$

Reduction of carboxylic acids

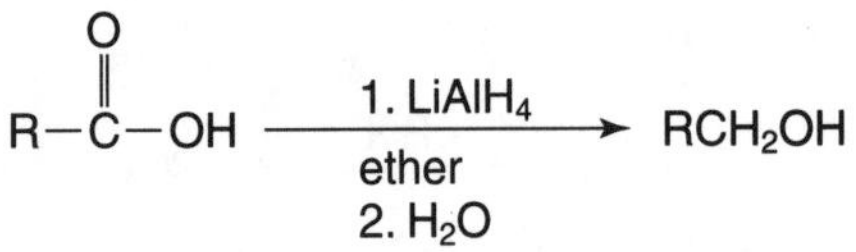

$$R-\overset{\overset{\displaystyle O}{\|}}{C}-OH \xrightarrow[\substack{ether \\ 2.\ H_2O}]{1.\ LiAlH_4} RCH_2OH$$

Reduction of esters

$$R-\overset{\overset{\displaystyle O}{\|}}{C}-OR' \xrightarrow[\substack{ether \\ 2.\ H_2O}]{1.\ LiAlH_4} R-CH_2OH \ + \ R'OH$$

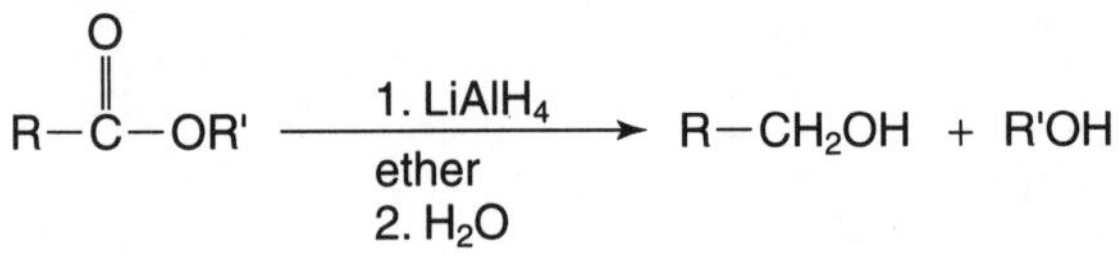

Grignard reagent with aldehydes and ketones

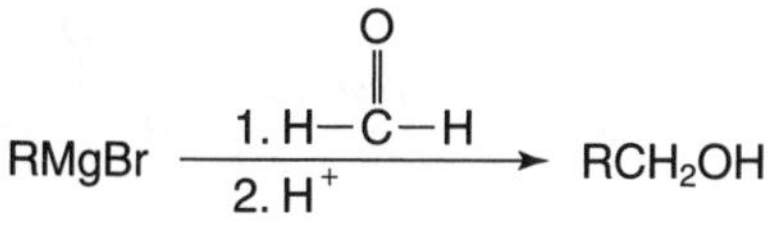

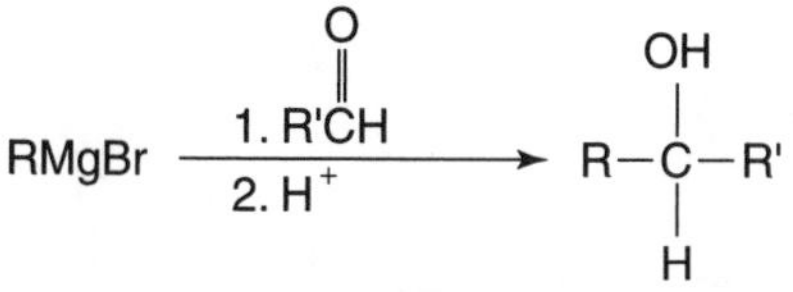

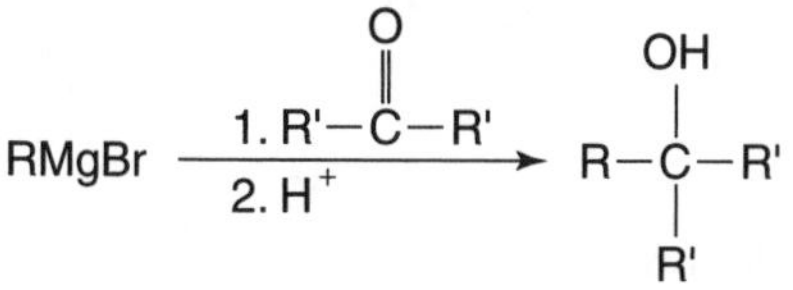

$$RMgBr \xrightarrow[\text{2. H}^+]{\text{1. } H-\overset{\displaystyle O}{\overset{\|}{C}}-H} RCH_2OH$$

$$RMgBr \xrightarrow[\text{2. H}^+]{\text{1. } R'\overset{\displaystyle O}{\overset{\|}{C}}H} R-\overset{\displaystyle OH}{\underset{\displaystyle H}{\overset{|}{\underset{|}{C}}}}-R'$$

$$RMgBr \xrightarrow[\text{2. H}^+]{\text{1. } R'-\overset{\displaystyle O}{\overset{\|}{C}}-R'} R-\overset{\displaystyle OH}{\underset{\displaystyle R'}{\overset{|}{\underset{|}{C}}}}-R'$$

Aldehydes

Oxidation of primary alcohols

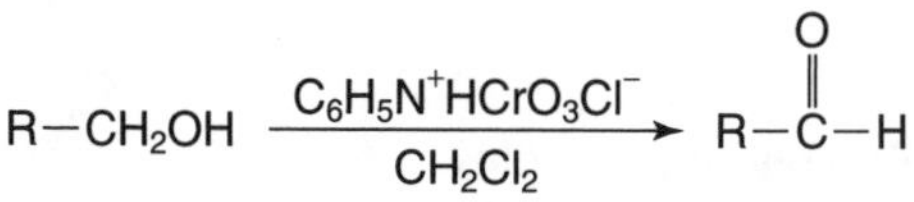

$$R-CH_2OH \xrightarrow[\text{CH}_2\text{Cl}_2]{\text{C}_6\text{H}_5\text{N}^+\text{HCrO}_3\text{Cl}^-} R-\overset{\displaystyle O}{\overset{\|}{C}}-H$$

Reduction of acyl halides

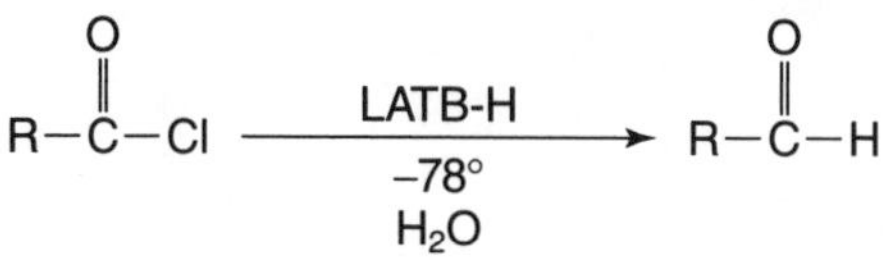

$$R-\overset{\displaystyle O}{\overset{\|}{C}}-Cl \xrightarrow[\substack{-78° \\ \text{H}_2\text{O}}]{\text{LATB-H}} R-\overset{\displaystyle O}{\overset{\|}{C}}-H$$

Reduction of esters

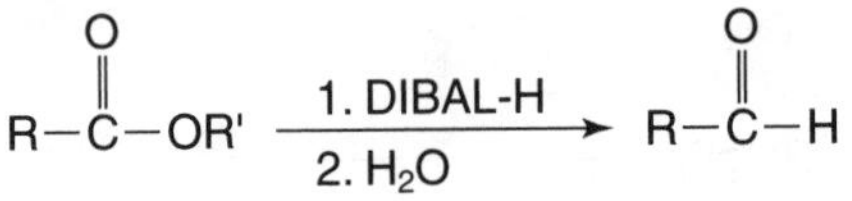

$$R-\overset{\overset{\textstyle O}{\|}}{C}-OR' \xrightarrow[\text{2. } H_2O]{\text{1. DIBAL-H}} R-\overset{\overset{\textstyle O}{\|}}{C}-H$$

Reduction of nitriles

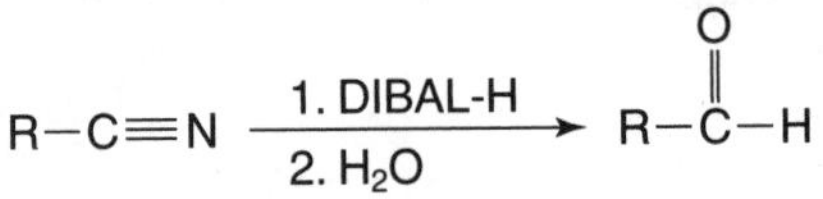

$$R-C\equiv N \xrightarrow[\text{2. } H_2O]{\text{1. DIBAL-H}} R-\overset{\overset{\textstyle O}{\|}}{C}-H$$

Ozonolysis of alkenes

$$R-CH=CH_2 \xrightarrow[H^+]{O_3} R-\overset{\overset{\textstyle O}{\|}}{C}-H \ + \ H-\overset{\overset{\textstyle O}{\|}}{C}-H$$

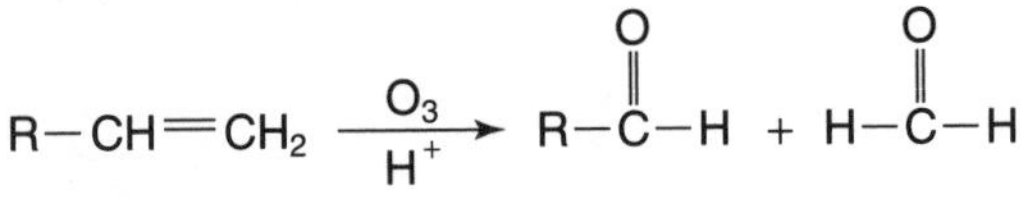

Ketones

Oxidation of secondary alcohols

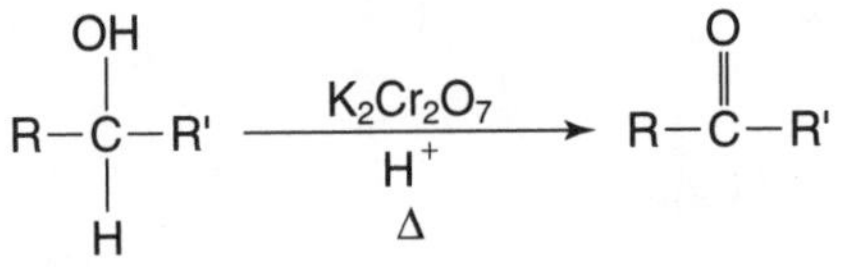

$$R-\overset{\overset{\textstyle OH}{|}}{\underset{\underset{\textstyle H}{|}}{C}}-R' \xrightarrow[\substack{H^+ \\ \Delta}]{K_2Cr_2O_7} R-\overset{\overset{\textstyle O}{\|}}{C}-R'$$

Hydration of alkynes

$$R-C\equiv C-H \xrightarrow[H^+]{H_2O} R-\overset{\overset{\displaystyle O}{\|}}{C}-CH_3$$

Ozonolysis of alkenes

$$R-\underset{\underset{\displaystyle R}{|}}{C}=CHR' \xrightarrow[2.\,H^+]{1.\,O_2} R-\overset{\overset{\displaystyle O}{\|}}{C}-R \;+\; R-\overset{\overset{\displaystyle O}{\|}}{C}-H$$

Friedel-Crafts acylation

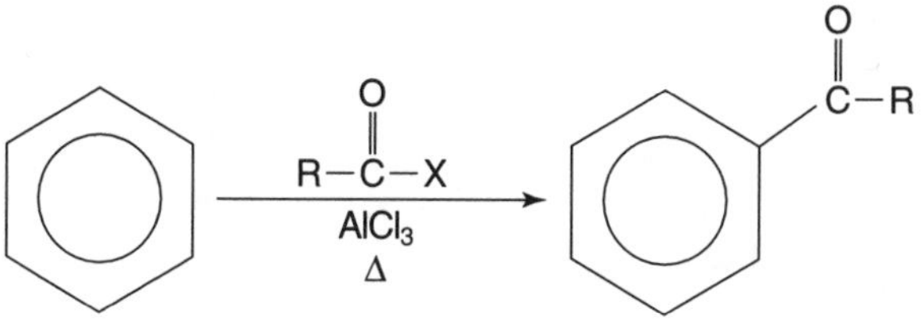

(used to prepare aromatic ketones only)

Via lithium dialkylcuprates

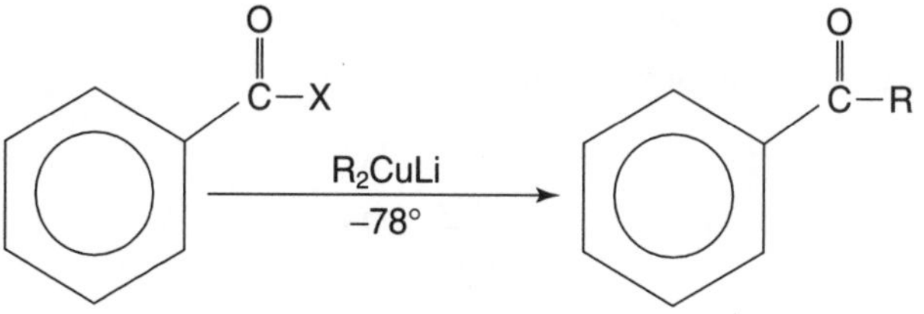

Via a Grignard reagent

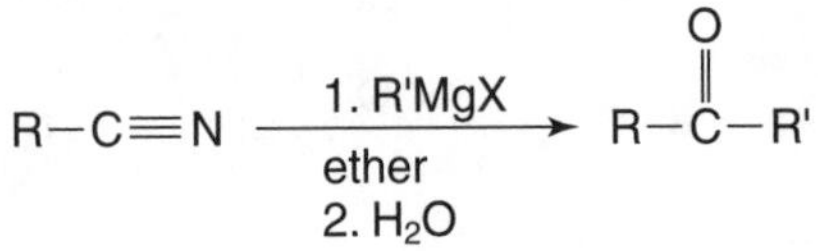

$$R-C\equiv N \quad \xrightarrow[\substack{\text{ether} \\ \text{2. H}_2\text{O}}]{\text{1. R'MgX}} \quad R-\overset{\displaystyle O}{\overset{\|}{C}}-R'$$

Carboxylic Acids

Oxidation of alkenes

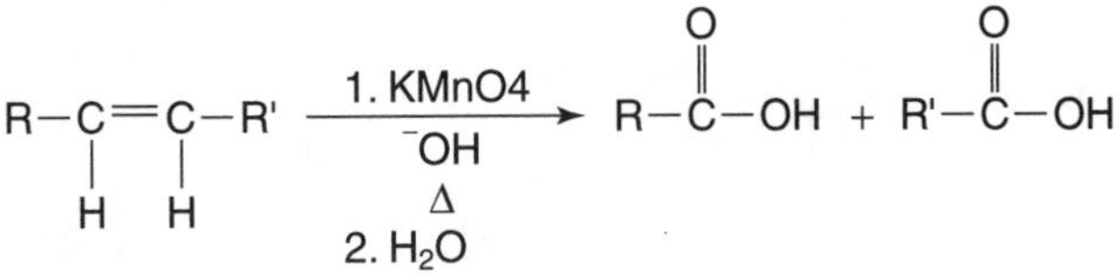

$$\underset{\substack{| \\ H}}{R-C}=\underset{\substack{| \\ H}}{C-R'} \quad \xrightarrow[\substack{\text{-OH} \\ \Delta \\ \text{2. H}_2\text{O}}]{\text{1. KMnO4}} \quad R-\overset{\displaystyle O}{\overset{\|}{C}}-OH \; + \; R'-\overset{\displaystyle O}{\overset{\|}{C}}-OH$$

Ozonolysis of alkenes

$$R-CH=CH-R' \quad \xrightarrow[\substack{\text{2. } R-\overset{\|}{\underset{O}{C}}-O-OH}]{\text{1. O}_3} \quad R-\overset{\displaystyle O}{\overset{\|}{C}}-OH \; + \; R'-\overset{\displaystyle O}{\overset{\|}{C}}-OH$$

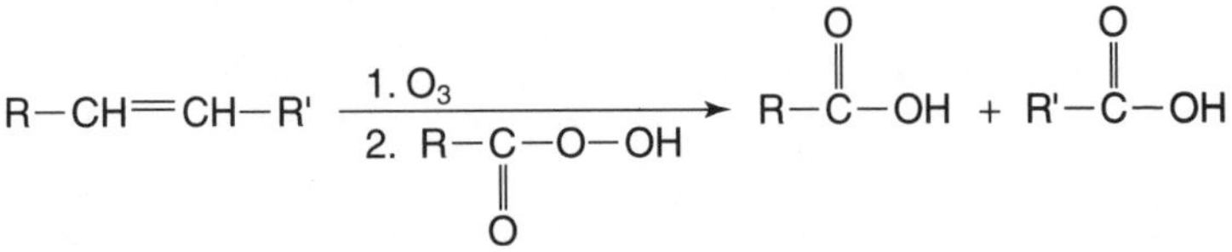

Oxidation of primary alcohols

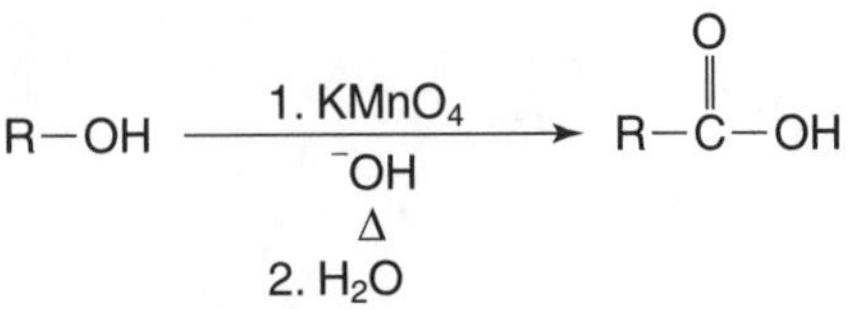

$$R-OH \quad \xrightarrow[\substack{\text{-OH} \\ \Delta \\ \text{2. H}_2\text{O}}]{\text{1. KMnO}_4} \quad R-\overset{\displaystyle O}{\overset{\|}{C}}-OH$$

Oxidation of aldehydes

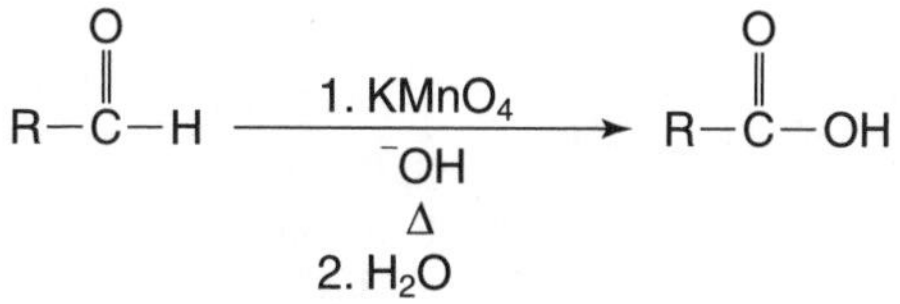

$$R-\overset{\displaystyle O}{\overset{\|}{C}}-H \xrightarrow[\substack{^-OH \\ \Delta \\ 2.\ H_2O}]{1.\ KMnO_4} R-\overset{\displaystyle O}{\overset{\|}{C}}-OH$$

Oxidation of alkyl benzenes

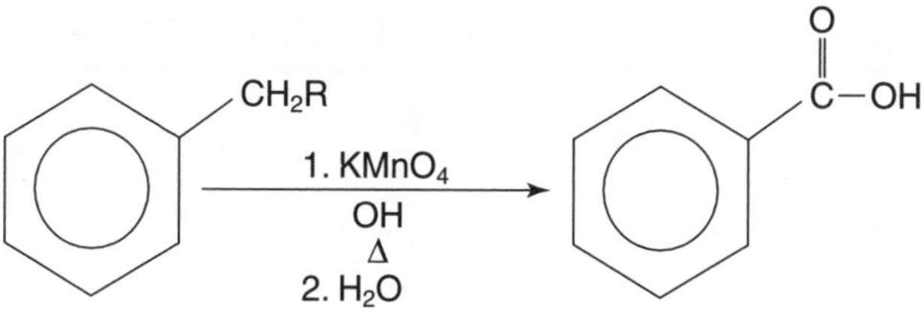

Hydrolysis of nitriles

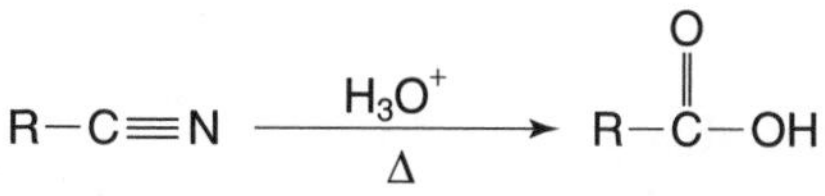

$$R-C\equiv N \xrightarrow[\Delta]{H_3O^+} R-\overset{\displaystyle O}{\overset{\|}{C}}-OH$$

Carbonation of a Grignard

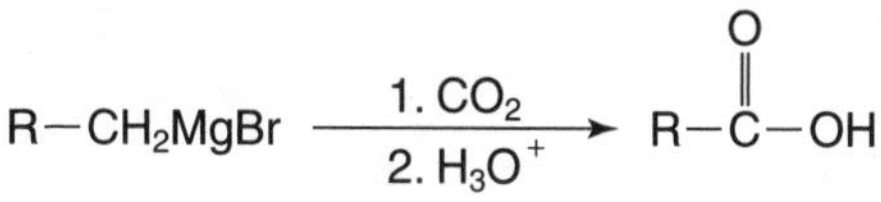

$$R-CH_2MgBr \xrightarrow[2.\ H_3O^+]{1.\ CO_2} R-\overset{\displaystyle O}{\overset{\|}{C}}-OH$$

Via acetoacetic ester

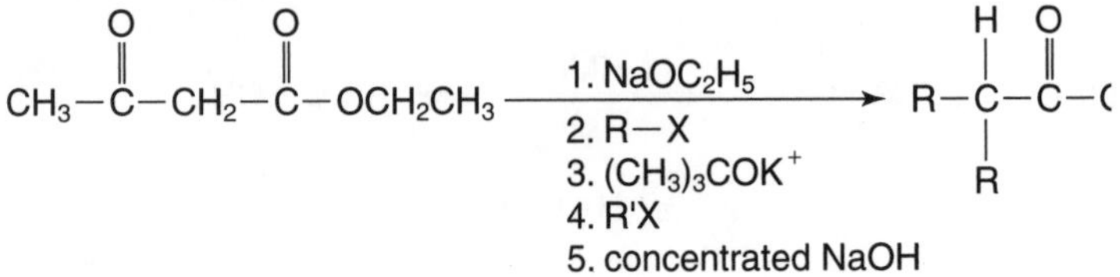

$$CH_3-\overset{\overset{\displaystyle O}{\|}}{C}-CH_2-\overset{\overset{\displaystyle O}{\|}}{C}-OCH_2CH_3 \xrightarrow[\substack{2.\ R-X \\ 3.\ (CH_3)_3COK^+ \\ 4.\ R'X \\ 5.\ \text{concentrated NaOH}}]{1.\ NaOC_2H_5} R-\overset{\overset{\displaystyle H}{|}}{\underset{\displaystyle R}{C}}-\overset{\overset{\displaystyle O}{\|}}{C}-($$

Halo Acids, α-Hydroxy Acids, and α, β-Unsaturated Acids

Hell-Volhard-Zelinski reaction

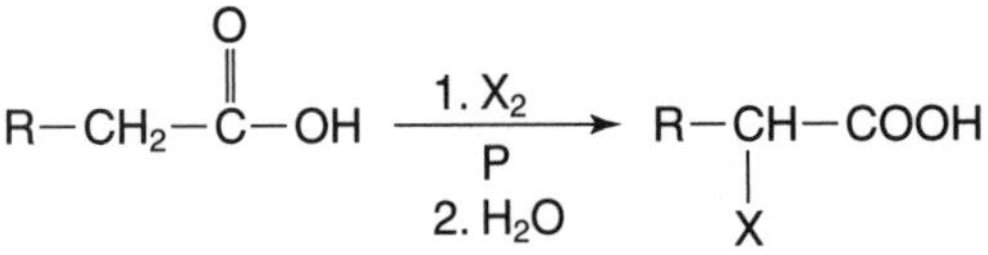

$$R-CH_2-\overset{\overset{\displaystyle O}{\|}}{C}-OH \xrightarrow[\substack{P \\ 2.\ H_2O}]{1.\ X_2} R-\underset{\underset{\displaystyle X}{|}}{CH}-COOH$$

Formation of α hydroxy acids

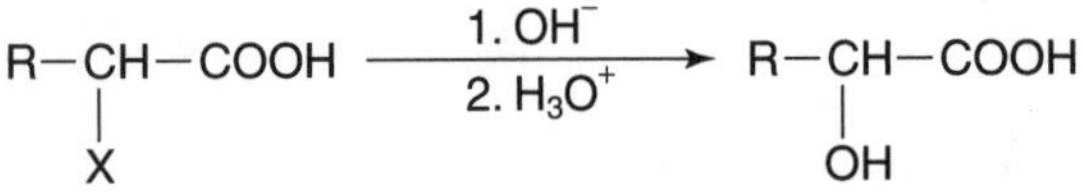

$$R-\underset{\underset{\displaystyle X}{|}}{CH}-COOH \xrightarrow[2.\ H_3O^+]{1.\ OH^-} R-\underset{\underset{\displaystyle OH}{|}}{CH}-COOH$$

Formation of amino acids

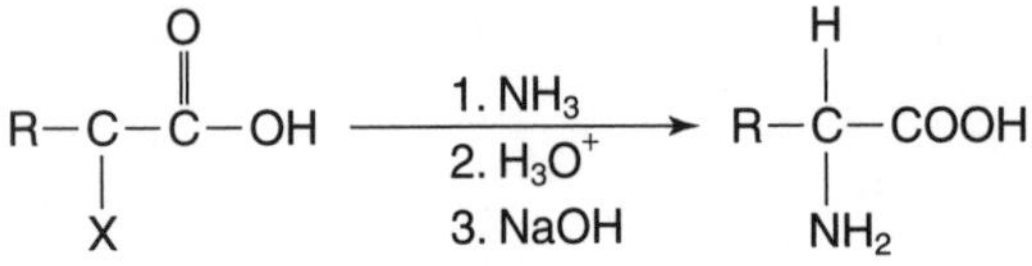

$$R-\overset{}{\underset{\underset{\displaystyle X}{|}}{C}}-\overset{\overset{\displaystyle O}{\|}}{C}-OH \xrightarrow[\substack{2.\ H_3O^+ \\ 3.\ NaOH}]{1.\ NH_3} R-\underset{\underset{\displaystyle NH_2}{|}}{\overset{\overset{\displaystyle H}{|}}{C}}-COOH$$

Formation of α, β-unsaturated acids

$$RCH_2\underset{\underset{X}{|}}{CH}-\overset{\overset{O}{\|}}{C}-OH \xrightarrow[\text{Alcohol}]{\text{KOH}} RCH=CH-\overset{\overset{O}{\|}}{C}-OH$$

Amines

Alkylation of ammonia

$$R-X \xrightarrow[\text{2. }^-OH]{\overset{\text{1. }NH_3}{\underset{\Delta}{}}} RNH_2$$

Via alkylazide reduction

$$R-N_3 \xrightarrow[\text{Alcohol}]{\text{Na}} R-NH_2$$

Gabriel synthesis

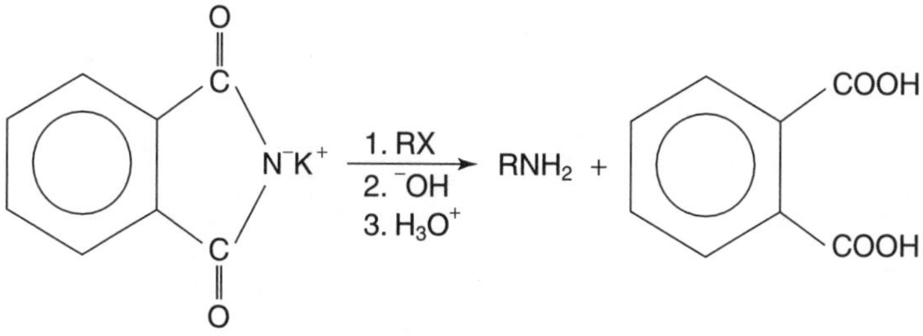

Reduction of nitriles

$$R-C\equiv N \xrightarrow[\substack{\text{ether} \\ \text{2. } H_3O^+}]{\text{1. } LiAlH_4} R-CH_2NH_2$$

Reduction of amides

$$R-\overset{\overset{\displaystyle O}{\|}}{C}-NH_2 \xrightarrow[\substack{\text{ether} \\ \text{2. } H_3O^+}]{\text{1. } LiAlH_4} RCH_2NH_2$$

Reduction of nitrocompounds

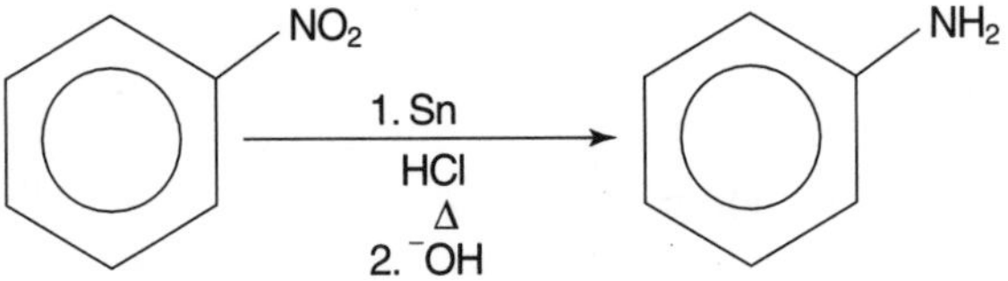

(used to prepare aromatic amines only)

Reductive amination of aldehydes and ketones

$$R-\overset{\overset{\displaystyle O}{\|}}{C}-R \quad \xrightarrow[\text{2. H}_2/\text{Ni}]{\text{1. NH}_2} \quad R-\overset{\overset{\displaystyle H}{|}}{\underset{\underset{\displaystyle H\ (R)}{|}}{C}}-NH_2$$

$$R-\overset{\overset{\displaystyle O}{\|}}{C}-H\ (R) \quad \xrightarrow[\text{2. H}_2/\text{Ni}]{\text{1. R'NH}_2} \quad R-\overset{\overset{\displaystyle H}{|}}{\underset{\underset{\displaystyle H\ (R)}{|}}{C}}-NHR$$

$$R-\overset{\overset{\displaystyle O}{\|}}{C}-H\ (R) \quad \xrightarrow[\text{2. H}_2/\text{Ni}]{\text{1. R'}_2\text{NH}} \quad R-\overset{\overset{\displaystyle H}{|}}{\underset{\underset{\displaystyle H\ (R)}{|}}{C}}-NR'_2$$

Aromatic Compounds

Halogenation

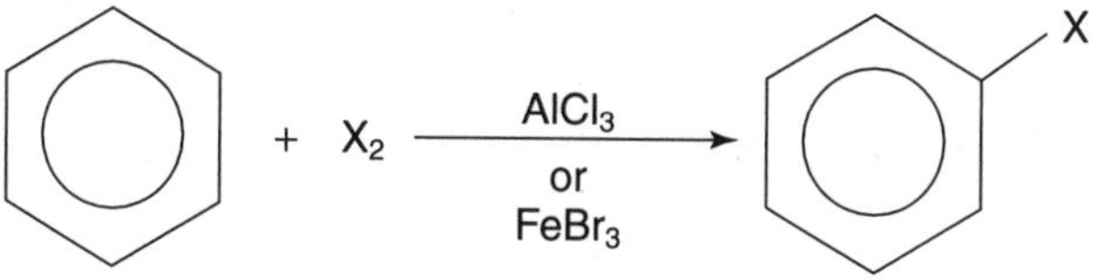

$$\text{C}_6\text{H}_6 + X_2 \xrightarrow[\text{or}\ FeBr_3]{AlCl_3} \text{C}_6\text{H}_5X$$

Nitration

$$\text{C}_6\text{H}_6 + HNO_3 + H_2SO_4 \xrightarrow{50°} \text{C}_6\text{H}_5NO_2$$

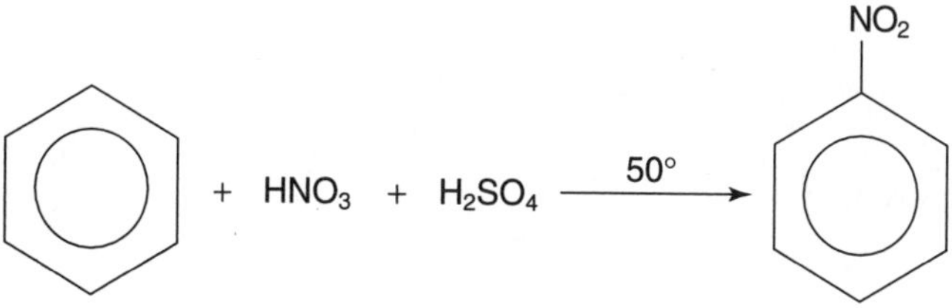

Sulfonation

$$\text{C}_6\text{H}_6 + H_2SO_4 \xrightarrow{25°} \text{C}_6\text{H}_5SO_3H$$

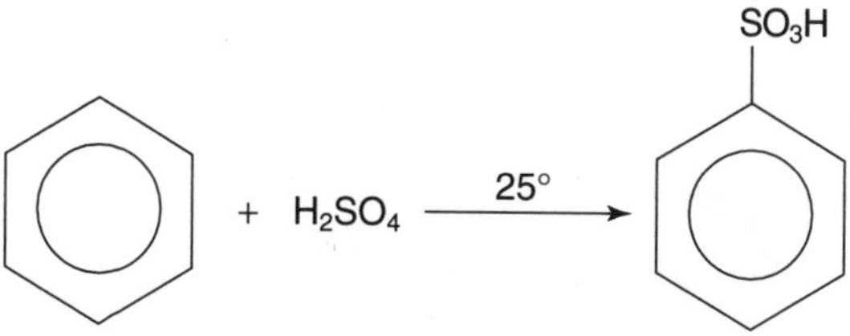

Friedel-Crafts alkylation

$$\text{benzene} \quad + \quad RX \quad \xrightarrow{\text{AlCl}_3} \quad \text{benzene}-R$$

(rearrangement of the R group may occur)

Friedel-Crafts acylation

$$\text{benzene} \quad + \quad R-\overset{\overset{\displaystyle O}{\|}}{C}-X \quad \xrightarrow{\text{AlCl}_3} \quad \text{benzene}-\overset{\overset{\displaystyle O}{\|}}{C}-R$$

Birch reduction

$$\text{benzene} \quad \xrightarrow[\substack{\text{NH}_3 \\ \text{CH}_3\text{OH}}]{\text{Na}} \quad \text{cyclohexadiene}$$

Alkyl Halides

Hydrolysis

$$R-X \ + \ NaOH \ \xrightarrow[\triangle]{H_2O} \ R-OH \ + \ NaX$$

Williamson ether synthesis

$$R-X \ \xrightarrow{R'O^-Na^+} \ R-OR' \ + \ NAX$$

Nitrile formation

$$R-X \ + \ NaCN \ \xrightarrow[\triangle]{H_2O} \ R-C\equiv N \ + \ NaX$$

Amine formation

$$R-X \ + \ NH_3 \ \xrightarrow[2. \ ^-OH]{1. \ \triangle} \ R-NH_2 \ + \ H_2O$$

$$R-X \ + \ R'NH_2 \ \xrightarrow[2. \ ^-OH]{1. \ \triangle} \ R-NHR' \ + \ H_2O$$

$$R-X \ + \ R_2'NH \ \xrightarrow[2. \ ^-OH]{1. \ \triangle} \ R-NHR_2' \ + \ H_2O$$

Alkene formation

$$CH_3CH_2X \ + \ RO^-K^+ \ \xrightarrow{\ \triangle\ } \ CH_2{=}CH_2 \ + \ K^+X^-$$

Grignard formation

$$R{-}X \ \xrightarrow[\text{ether}]{\text{Mg}} \ R{-}Mg{-}X$$

Phenols

Neutralization

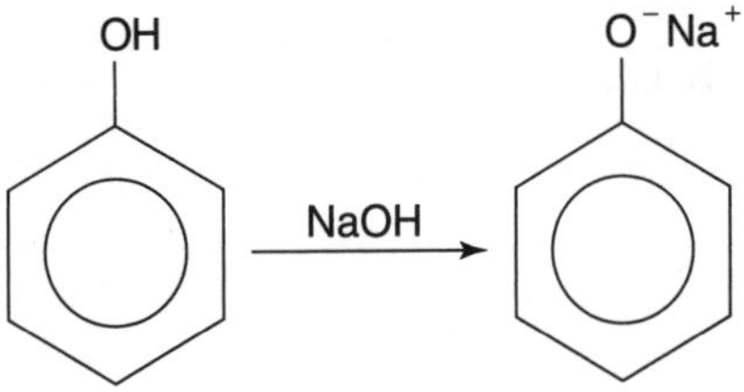

Ester formation

Ether formation

Halogenation

Nitration

Sulfonation

Kolbe reaction

Aryl Halides

Grignard reaction

Hydrolysis

Amination

Ethers

Cleavage

$$R-O-R' \xrightarrow{\text{HI}} R-OH + R'I$$

Protonation

$$R-\ddot{O}-R' \xrightarrow{\text{H}^+} R-\overset{\text{H}}{\underset{+}{\ddot{O}}}-R'$$

Alcohols

Neutralization

$$R-OH \xrightarrow{\text{Na}} RO^- Na^+ + H_2$$

Halide Formation

$$R-OH \xrightarrow[\triangle]{\text{HX}} RX$$

$$ROH \xrightarrow[\triangle]{\text{SOX}_2} RX$$

$$ROH \xrightarrow[\triangle]{\text{PCl}_3} R-Cl$$

$$ROH \xrightarrow[\triangle]{\text{PCl}_5} RCl$$

Ester formation

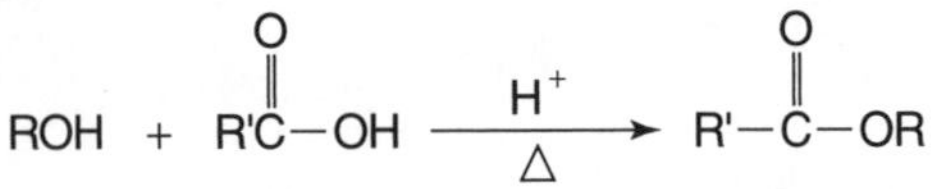

$$ROH \ + \ R'\overset{\displaystyle O}{\underset{\displaystyle \|}{C}}{-}OH \ \xrightarrow[\Delta]{H^+} \ R'{-}\overset{\displaystyle O}{\underset{\displaystyle \|}{C}}{-}OR$$

Oxidation

$$R{-}OH \ \xrightarrow[\substack{reagent \\ 25°}]{Sarett's} \ R{-}\overset{\displaystyle O}{\underset{\displaystyle \|}{C}}{-}H$$

$$R{-}\overset{\displaystyle OH}{\underset{\displaystyle OH}{\overset{\displaystyle |}{\underset{\displaystyle |}{C}}R'}} \ \xrightarrow[\substack{H^+ \\ 70°}]{K_2Cr_2O_7} \ R{-}\overset{\displaystyle O}{\underset{\displaystyle \|}{C}}{-}R'$$

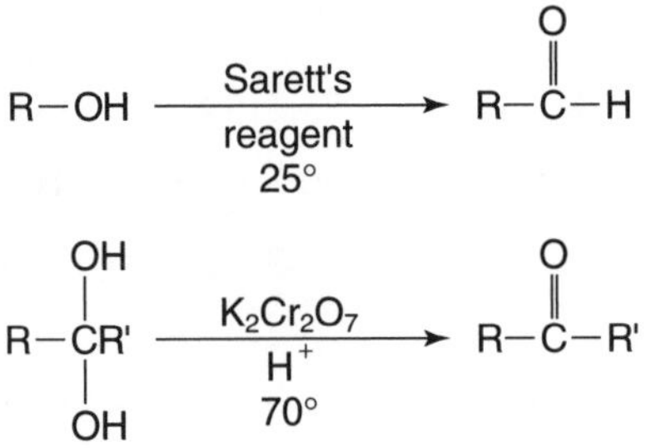

Carboxylic acid formation

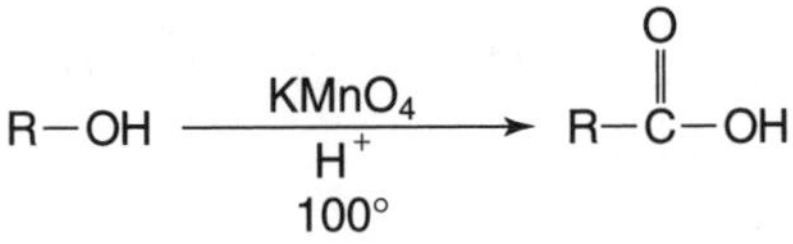

$$R{-}OH \ \xrightarrow[\substack{H^+ \\ 100°}]{KMnO_4} \ R{-}\overset{\displaystyle O}{\underset{\displaystyle \|}{C}}{-}OH$$

Aldehydes and Ketones

Hydration

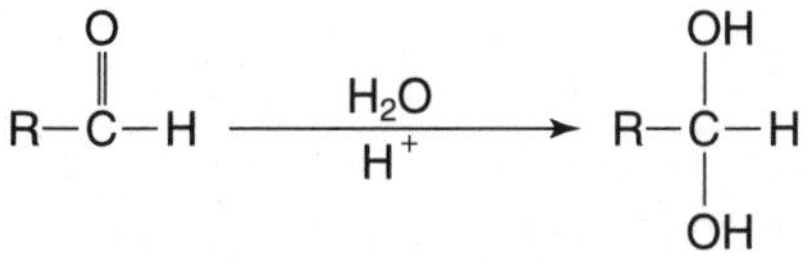

Hemiacetal and acetal formation

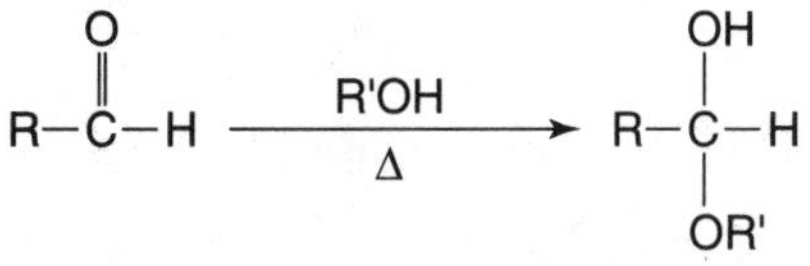

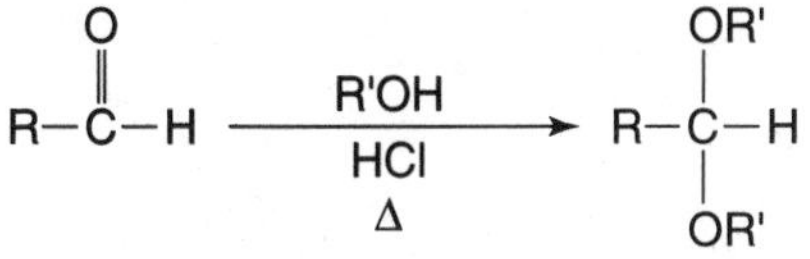

Hydrogen cyanide addition

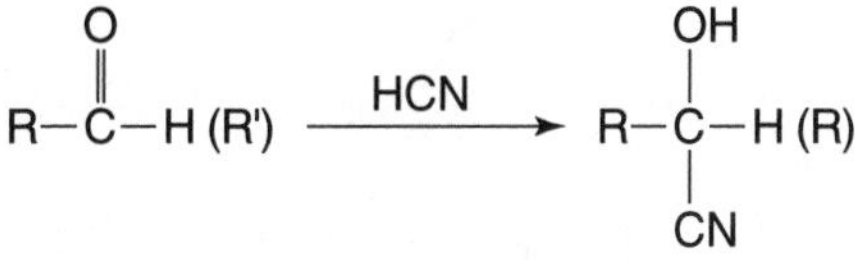

Addition of organometallic reagents

Addition of ammonia derivatives

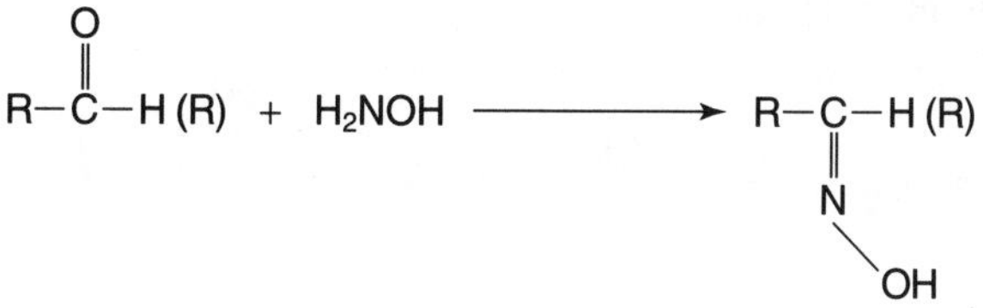

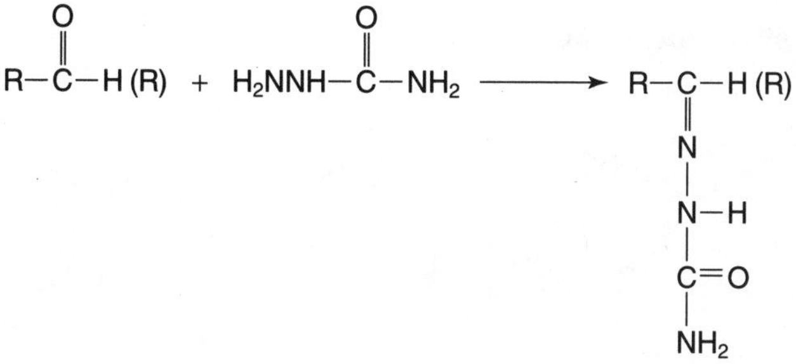

$$R-\overset{\overset{\displaystyle O}{\|}}{C}-H\,(R)\;+\;H_2NOH\;\longrightarrow\;R-\overset{\overset{\displaystyle \|}{N}}{C}-H\,(R)$$

with N—OH on the product

$$R-\overset{\overset{\displaystyle O}{\|}}{C}-H\,(R)\;+\;H_2NNH-\overset{\overset{\displaystyle O}{\|}}{C}-NH_2\;\longrightarrow\;R-\overset{\displaystyle \|}{C}-H\,(R)$$

N
N—H
C=O
NH₂

Oxidation of aldehydes

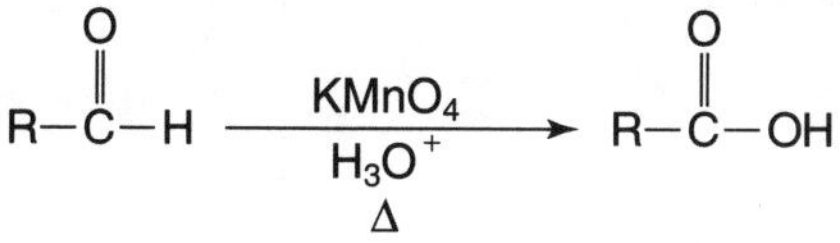

$$R-\overset{\overset{\displaystyle O}{\|}}{C}-H\;\xrightarrow[\Delta]{\underset{H_3O^+}{KMnO_4}}\;R-\overset{\overset{\displaystyle O}{\|}}{C}-OH$$

α-halogenation of ketones

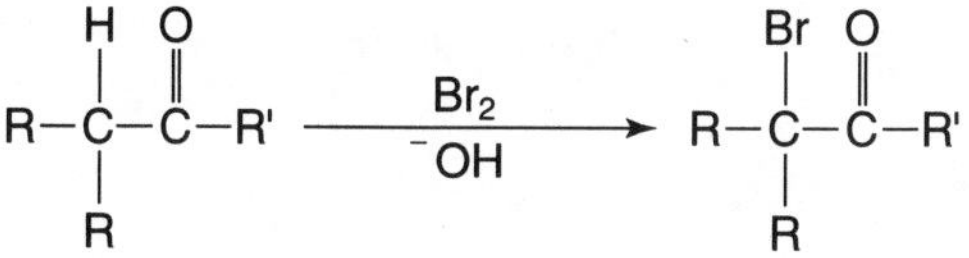

$$R-\overset{\overset{\displaystyle H}{|}}{\underset{\displaystyle R}{C}}-\overset{\overset{\displaystyle O}{\|}}{C}-R'\;\xrightarrow[^-OH]{Br_2}\;R-\overset{\overset{\displaystyle Br}{|}}{\underset{\displaystyle R}{C}}-\overset{\overset{\displaystyle O}{\|}}{C}-R'$$

Aldol condensation

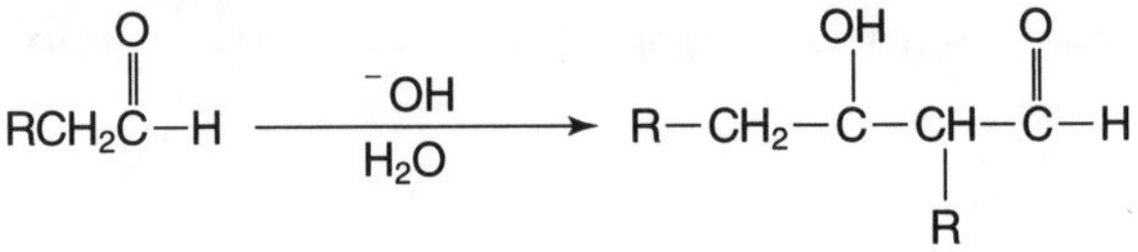

$$RCH_2\overset{\displaystyle O}{\overset{\displaystyle \|}{C}}-H \xrightarrow[\text{H}_2\text{O}]{^-\text{OH}} R-CH_2-\underset{\displaystyle R}{\overset{\displaystyle OH}{\underset{\displaystyle |}{\overset{\displaystyle |}{C}}}}-CH-\overset{\displaystyle O}{\overset{\displaystyle \|}{C}}-H$$

Benzoin condensation

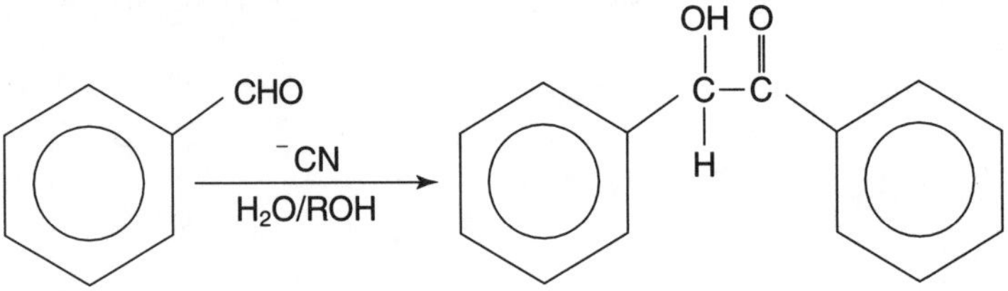

Carboxylic Acids

Ester formation

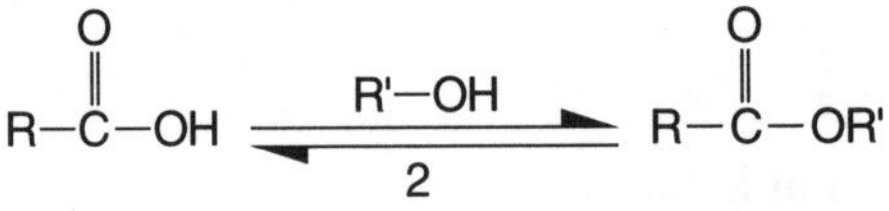

$$R-\overset{\displaystyle O}{\overset{\displaystyle \|}{C}}-OH \underset{2}{\overset{R'-OH}{\rightleftharpoons}} R-\overset{\displaystyle O}{\overset{\displaystyle \|}{C}}-OR'$$

(Fischer Method)

Nonreversible ester formation

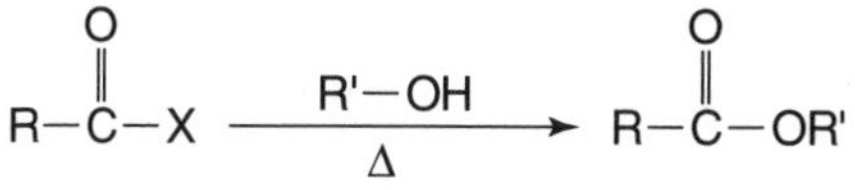

$$R-\overset{\overset{\displaystyle O}{\|}}{C}-X \quad \xrightarrow[\Delta]{R'-OH} \quad R-\overset{\overset{\displaystyle O}{\|}}{C}-OR'$$

Methyl ketone formation

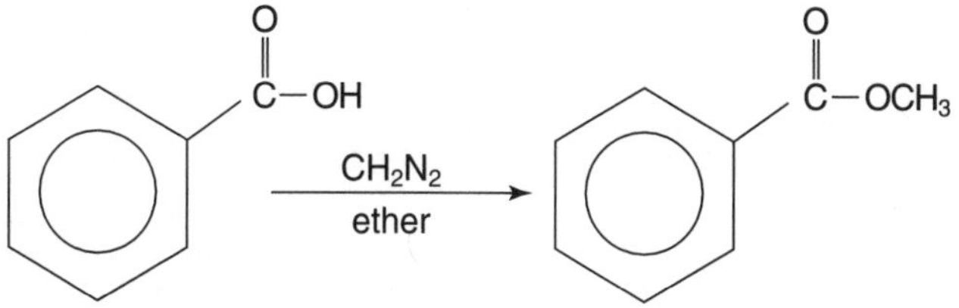

Amide formation

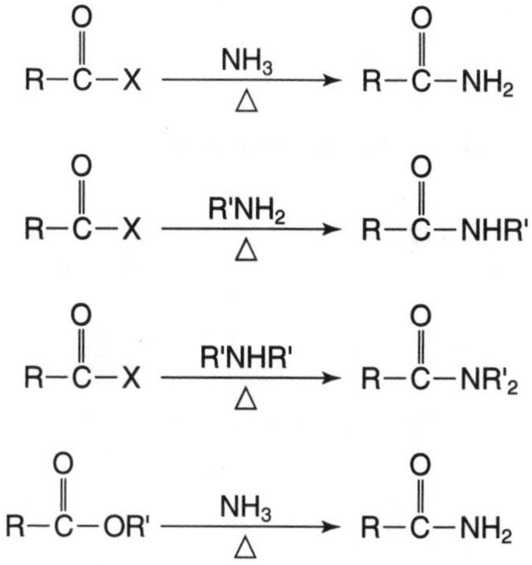

$$R-\overset{\overset{\displaystyle O}{\|}}{C}-X \quad \xrightarrow[\Delta]{NH_3} \quad R-\overset{\overset{\displaystyle O}{\|}}{C}-NH_2$$

$$R-\overset{\overset{\displaystyle O}{\|}}{C}-X \quad \xrightarrow[\Delta]{R'NH_2} \quad R-\overset{\overset{\displaystyle O}{\|}}{C}-NHR'$$

$$R-\overset{\overset{\displaystyle O}{\|}}{C}-X \quad \xrightarrow[\Delta]{R'NHR'} \quad R-\overset{\overset{\displaystyle O}{\|}}{C}-NR'_2$$

$$R-\overset{\overset{\displaystyle O}{\|}}{C}-OR' \quad \xrightarrow[\Delta]{NH_3} \quad R-\overset{\overset{\displaystyle O}{\|}}{C}-NH_2$$

Acid halide formation

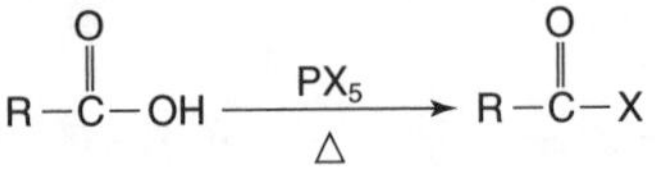

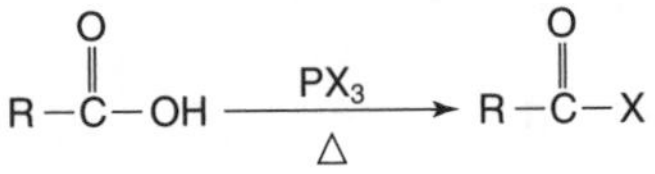

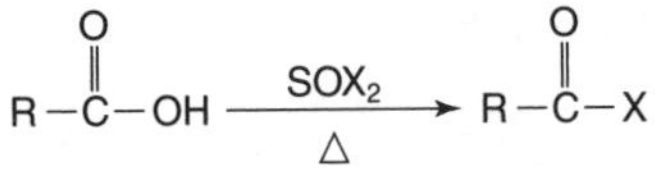

$$R-\overset{\overset{\displaystyle O}{\|}}{C}-OH \xrightarrow[\triangle]{PX_5} R-\overset{\overset{\displaystyle O}{\|}}{C}-X$$

$$R-\overset{\overset{\displaystyle O}{\|}}{C}-OH \xrightarrow[\triangle]{PX_3} R-\overset{\overset{\displaystyle O}{\|}}{C}-X$$

$$R-\overset{\overset{\displaystyle O}{\|}}{C}-OH \xrightarrow[\triangle]{SOX_2} R-\overset{\overset{\displaystyle O}{\|}}{C}-X$$

Anhydride formation

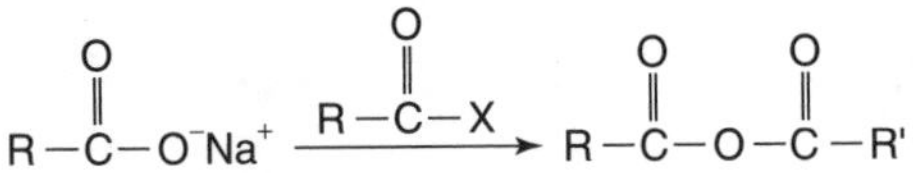

$$R-\overset{\overset{\displaystyle O}{\|}}{C}-O^-Na^+ \xrightarrow{R-\overset{\overset{\displaystyle O}{\|}}{C}-X} R-\overset{\overset{\displaystyle O}{\|}}{C}-O-\overset{\overset{\displaystyle O}{\|}}{C}-R'$$

Decarboxylation-Hunsdiecker reaction

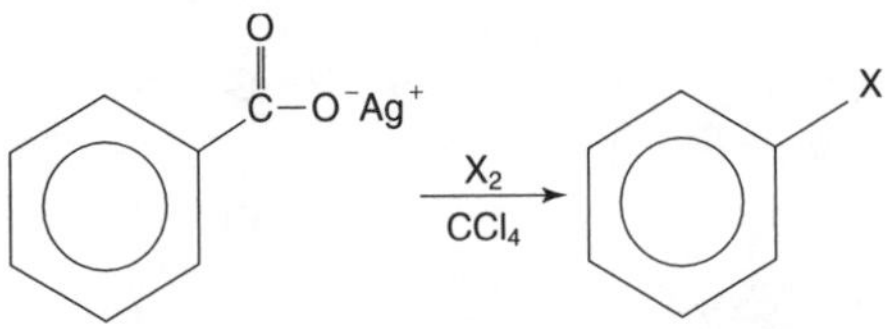

(must be silver salt of aromatic carboxylic acid)

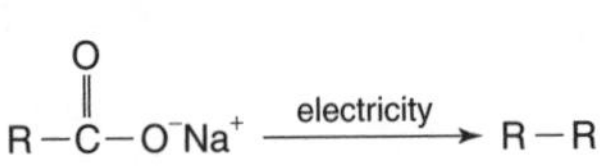

$$R-\overset{\overset{\displaystyle O}{\|}}{C}-O^-Na^+ \xrightarrow{electricity} R-R$$

(Kolbe electrolysis)

Decarboxylation with copper salts

$$R-\overset{\overset{\displaystyle O}{\|}}{C}-OH \xrightarrow[\Delta]{\underset{\text{quinoline}}{\text{Cu}}} RH \ + \ CO_2$$

Reduction of carboxylic acids

$$R-\overset{\overset{\displaystyle O}{\|}}{C}-OH \xrightarrow[\substack{\text{ether} \\ \text{2. } H_3O^+}]{\text{1. } LiAlH_4} RCH_2OH$$

Esters

Reduction

$$R-\overset{\overset{\displaystyle O}{\|}}{C}-OR' \xrightarrow[\substack{\text{ether} \\ \text{2. } H_3O^+}]{\text{1. } LiAlH_4} R-OH \ + \ R'OH$$

Acid halide

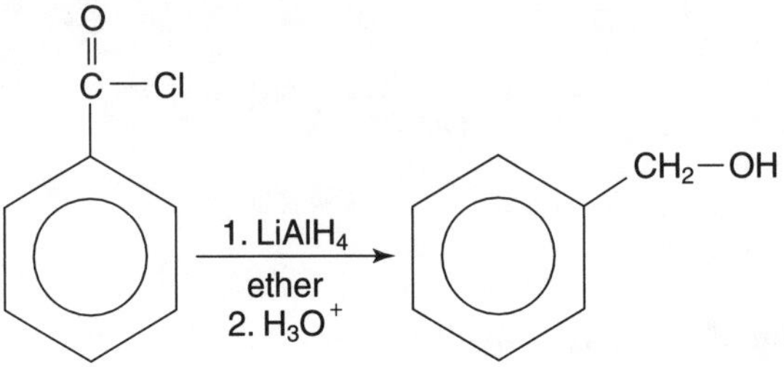

Acyl Halides

Esterification

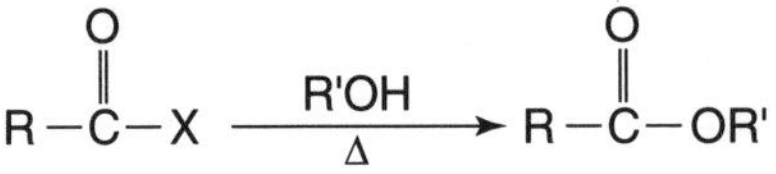

Anhydride formation

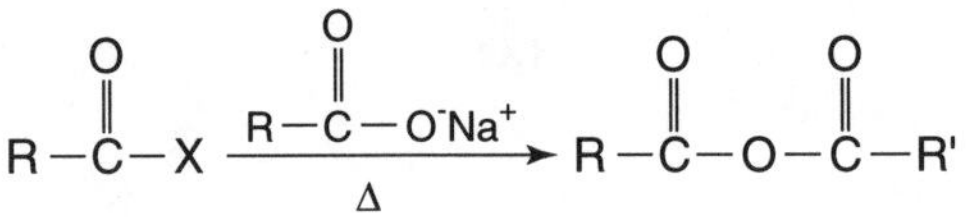

Amide formation

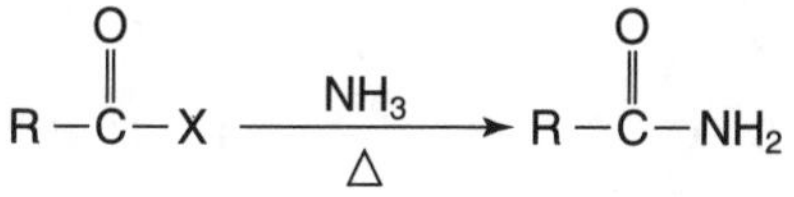

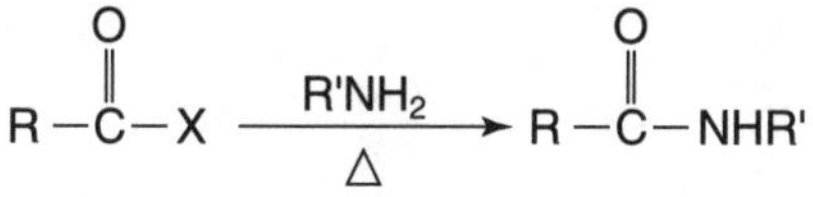

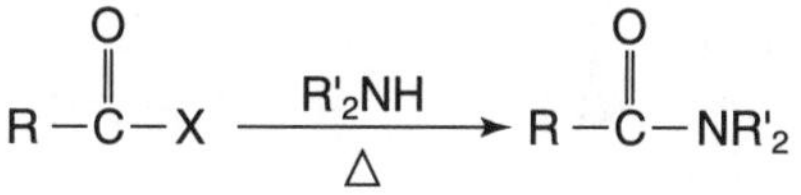

$$R-\overset{\overset{\displaystyle O}{\|}}{C}-X \xrightarrow[\triangle]{NH_3} R-\overset{\overset{\displaystyle O}{\|}}{C}-NH_2$$

$$R-\overset{\overset{\displaystyle O}{\|}}{C}-X \xrightarrow[\triangle]{R'NH_2} R-\overset{\overset{\displaystyle O}{\|}}{C}-NHR'$$

$$R-\overset{\overset{\displaystyle O}{\|}}{C}-X \xrightarrow[\triangle]{R'_2NH} R-\overset{\overset{\displaystyle O}{\|}}{C}-NR'_2$$

Acid formation

$$R-\overset{\overset{\displaystyle O}{\|}}{C}-X \xrightarrow[\triangle]{H_2O} R-\overset{\overset{\displaystyle O}{\|}}{C}-OH$$

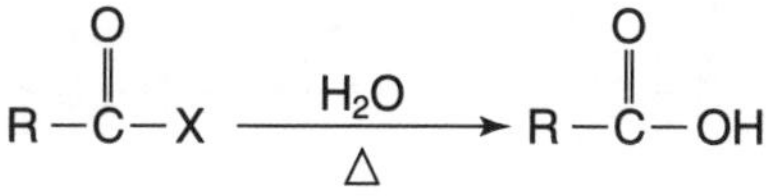

Anhydrides

Ester formation

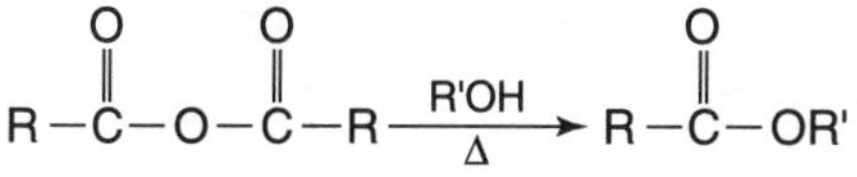

$$R-\overset{\overset{\displaystyle O}{\|}}{C}-O-\overset{\overset{\displaystyle O}{\|}}{C}-R \xrightarrow[\triangle]{R'OH} R-\overset{\overset{\displaystyle O}{\|}}{C}-OR'$$

Amide formation

$$R-\overset{\displaystyle O}{\overset{\|}{C}}-O-\overset{\displaystyle O}{\overset{\|}{C}}-R \xrightarrow[\triangle]{NH_3} R-\overset{\displaystyle O}{\overset{\|}{C}}-NH_2$$

$$R-\overset{\displaystyle O}{\overset{\|}{C}}-O-\overset{\displaystyle O}{\overset{\|}{C}}-R \xrightarrow[\triangle]{RNH_2} R-\overset{\displaystyle O}{\overset{\|}{C}}-NH(R)$$

$$R-\overset{\displaystyle O}{\overset{\|}{C}}-O-\overset{\displaystyle O}{\overset{\|}{C}}-R \xrightarrow[\triangle]{R_2NH} R-\overset{\displaystyle O}{\overset{\|}{C}}-NR_2$$

Acid formation

$$R-\overset{\displaystyle O}{\overset{\|}{C}}-O-\overset{\displaystyle O}{\overset{\|}{C}}-R \xrightarrow[\triangle]{H_2O} R-\overset{\displaystyle O}{\overset{\|}{C}}-OH$$

Amines

Ammonium ion formation

$$R-NH_2 \xrightarrow{HX} R\overset{+}{N}H_3X^-$$

Amide formation

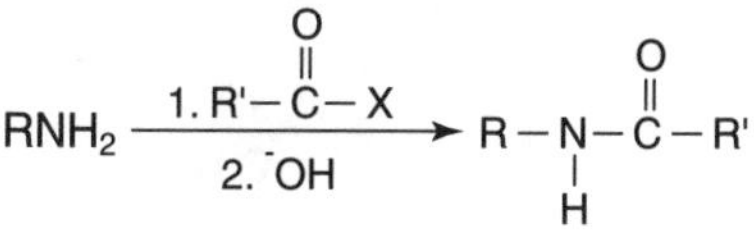

$$RNH_2 \xrightarrow[\text{2. }^-OH]{\text{1. } R'-\overset{\overset{O}{\|}}{C}-X} R-\overset{\underset{H}{|}}{N}-\overset{\overset{O}{\|}}{C}-R'$$

Benzene sulfonamide formation

$$RNH_2 \xrightarrow[\Delta]{C_6H_5SO_2Cl} R-NH-SO_2C_6H_5$$

Oxidation

$$R_3N \xrightarrow[\substack{\text{or} \\ \text{peroxyacetic} \\ \text{acid}}]{H_2O_2} R_3N^+O^-$$

Reaction with nitrous acid

$$RNH_2 \xrightarrow[0°]{\substack{NaNO_2 \\ HCl}} \left[RN_2{}^+Cl^- \right] \longrightarrow R^+ + N_2$$

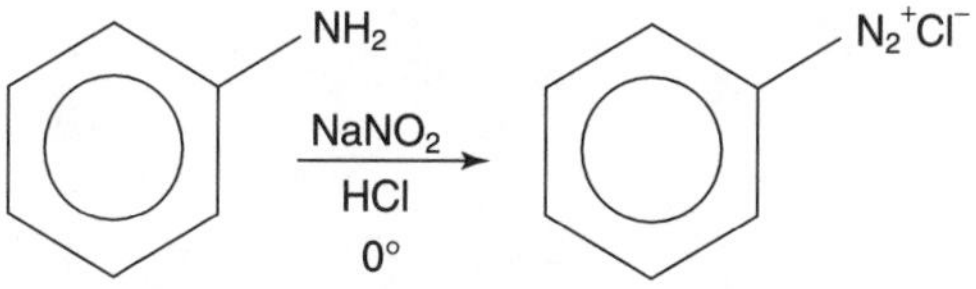

$$R_2NH \xrightarrow[HCl]{NaNO_2} R_2N-N=O$$

Reaction of diazonium salt

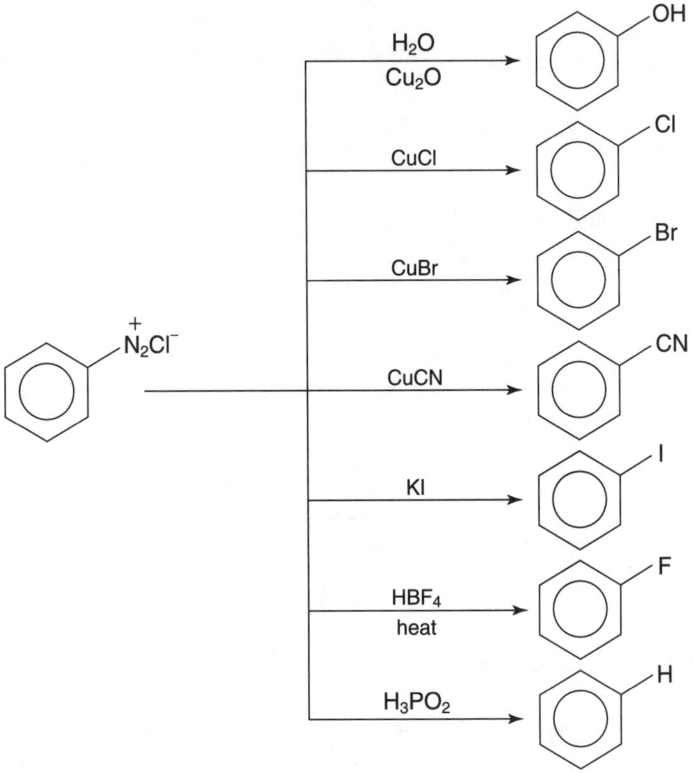

acetal the product formed by the reaction of an aldehyde with an alcohol. The general structure of an acetal is:

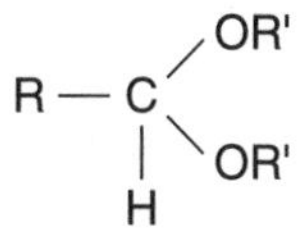

$$R - \overset{\displaystyle |}{\underset{\displaystyle H}{C}} \overset{\displaystyle OR'}{\underset{\displaystyle OR'}{<}}$$

achiral the opposite of **chiral;** also called *nonchiral.* An achiral molecule can be superimposed on its mirror image.

acid see **Brønsted-Lowry theory of acids and bases,** and **Lewis theory of acids and bases.**

acid-base reaction a neutralization reaction in which the products are a salt and water.

activated complex molecules at an unstable intermediate stage in a reaction.

activating group a group that increases the rate of electrophilic aromatic substitution when bonded to an aromatic ring.

activation energy the energy that must be supplied to chemicals to initiate a reaction; the difference in potential energy between the ground state and the transition state of molecules. Molecules of reactants must have this amount of energy to proceed to the product state.

acyl group a group with the following structure, where R can be either an alkyl or aryl group.

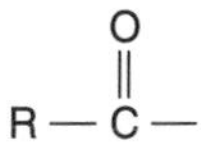

$$R - \overset{\displaystyle O}{\overset{\displaystyle \|}{C}} -$$

acyl halide a compound with the general structural formula:

$$\begin{array}{c} \quad\; O \\ \quad\; \| \\ R-C-X \end{array}$$

acylation a reaction in which an acyl group is added to a molecule.

acylium ion the resonance stabilized cation:

$$R\overset{+}{C}=\overset{..}{O}: \longleftrightarrow RC\equiv\overset{+}{O}:$$

addition a reaction that produces a new compound by combining all of the elements of the original reactants.

addition elimination mechanism the two-stage mechanism by which nucleophilic aromatic substitution occurs. In the first stage, addition of the nucleophile to the carbon bearing the leaving group occurs. An elimination follows in which the leaving group is expelled.

adduct the product of an addition reaction.

alcohol an organic chemical that contains an —OH group.

aldehyde an organic chemical that contains a —CHO group.

alicyclic compound an *ali*phatic *cyclic* hydrocarbon, which means that a compound contains a ring but not an aromatic ring.

aliphatic compound a straight- or branched-chain hydrocarbon; an alkane, alkene, or alkyne.

alkane a hydrocarbon that contains only single covalent bonds. The alkane general formula is C_nH_{2n+2}.

alkene a hydrocarbon that contains a carbon-carbon double bond. The alkene general formula is C_nH_{2n}.

alkoxide ion an anion formed by removing a proton from an alcohol; the RO^- ion.

alkoxy free radical a free radical formed by the homolytic cleavage of an alcohol —OH bond; the $RO\cdot$ free radical.

alkyl group an alkane molecule from which a hydrogen atom has been removed. Alkyl groups are abbreviated as "R" in structural formulas.

alkyl halide a hydrocarbon that contains a halogen substituent, such as fluorine, chlorine, bromine, or iodine.

alkyl-substituted cycloalkane a cyclic hydrocarbon to which one or more alkyl groups are bonded. (Compare with **cycloalkyl alkane.**)

alkylation a reaction in which an alkyl group is added to a molecule.

alkyne a hydrocarbon that contains a triple bond. The alkyne general formula is C_nH_{2n-2}.

allyl group the $H_2C{=}CHCH_2-$ group.

allylic carbocation the $H_2C{=}CHCH_2^+$ ion.

analogue in organic chemistry, chemicals that are similar to each other, but not identical. For example, the hydrocarbons are all similar to each other, but an alkane is different from the alkenes and alkynes because of the types of bonds they contain. Therefore, an alkane and an alkene are analogues.

angle of rotation (α) in a polarimeter, the angle right or left in which plane-polarized light is turned after passing through an optically active compound in solution.

anion a negatively charged ion.

antibonding molecular orbital a molecular orbital that contains more energy than the atomic orbitals from which it was formed; in other words, an electron is less stable in an antibonding orbital than it is in its original atomic orbital.

anti-Markovnikov addition a reaction in which the hydrogen atom of a hydrogen halide bonds to the carbon of a double bond that is bonded to *fewer* hydrogen atoms. The addition takes place via a free-radical intermediate rather than a carbocation. (Compare with **Markovnikov rule.**)

arene an aromatic hydrocarbon.

aromatic compound a compound that possesses a closed-shell electron configuration as well as resonance. This type of compound obeys Hückel's rule.

aryl group a group produced by the removal of a proton from an aromatic molecule.

aryl halide a compound in which a halogen atom is attached to an aromatic ring.

atom the smallest amount of an element; a nucleus surrounded by electrons.

atomic mass (A) the sum of the weights of the protons and neutrons in an atom. (A proton and neutron each have a mass of 1 atomic mass unit.)

atomic number (Z) the number of protons or electrons in an atom.

atomic 1s orbital the spherical orbital nearest the nucleus of an atom.

atomic orbital a region in space around the nucleus of an atom where the probability of finding an electron is high.

atomic *p* orbital an hourglass-shaped orbital, oriented on *x, y,* and *z* axes in three-dimensional space.

atomic *s* orbital a spherical orbital.

Baeyer reagent Cold, dilute potassium permanganate, which is used to oxidize alkenes and alkynes.

base see **Brønsted-Lowry theory of acids and bases,** and **Lewis theory of acids and bases.**

benzenoid ring an aromatic ring with a benzene-like structure.

benzyl group the $C_6H_5CH_2$ group.

benzyne an unstable intermediate that consists of a benzene ring with an additional bond that is created by the side-to-side overlap of sp^2 orbitals on adjacent carbons of the ring.

bond angle the angle formed between two adjacent bonds on the same atom.

bond-dissociation energy the amount of energy needed to homolytically fracture a bond.

bond length the equilibrium distance between the nuclei of two atoms or groups that are bonded to each other.

bond strength see **bond-dissociation energy**.

bonding electron see **valence electrons.**

bonding molecular orbital the orbital formed by the overlap of adjacent atomic orbitals.

branched-chain alkane an alkane with alkyl groups bonded to the central carbon chain.

Brønsted-Lowry theory of acids and bases A Brønsted-Lowry acid is a compound capable of donating a proton (a hydrogen ion), and a Brønsted-Lowry base is capable of accepting a hydrogen ion. In *neutralization,* an acid donates a proton to a base, creating a conjugate acid and a conjugate base.

carbanion a carbon atom bearing a negative charge; a carbon anion.

carbene an electrically uncharged molecule that contains a carbon atom with only two single bonds and just six electrons in its valence shell.

carbenoid a chemical that resembles a carbene in its chemical reactions.

carbocation a carbon cation; a carbon atom bearing a positive charge (sometimes referred to as a "carbonium ion").

carbonyl group the $-\overset{\overset{\displaystyle O}{\|}}{C}-$ group.

carboxylic acid the $-\underset{\underset{\displaystyle O}{\|}}{C}-OH$ group.

catalyst a substance that affects the rate of a reaction in which it participates; however, it is not altered or used up in the process.

cation a positively charged ion.

cationic polymerization occurs via a cation intermediate and is less efficient than free-radical polymerization.

chain reaction a reaction that, once started, produces sufficient energy to keep the reaction running. These reactions proceed by a series of steps, which produce intermediates, energy, and products.

chemical shift a position in an NMR spectrum, relative to TMS, at which a nucleus absorbs.

chiral describes a molecule that is not superimposable on its mirror image; like the relationship of a left hand to a right hand.

closed-shell electron configuration a stable electron configuration in which all of the electrons are located in the lowest energy orbitals available.

competing reactions two reactions that start with the same reactants but form different products.

concerted taking place at the same time without the formation of an intermediate.

condensation reaction a reaction in which two molecules join with the liberation of a small stable molecule.

conjugate acid the acid that results when a Brønsted-Lowry base accepts a hydrogen ion.

conjugate base the base that results when a Brønsted-Lowry acid loses a hydrogen ion.

conjugated double bonds carbon-carbon double bonds that are separated from one another by one single bond; $C\!=\!C\!-\!C\!=\!C$, for example.

conjugation the overlapping in all directions of a series of p orbitals. This process usually occurs in a molecule with alternating double and single bonds.

conjugation energy see **resonance energy.**

coupling constant (J) the separation in frequency units between multiple peaks in one chemical shift. This separation results from spin-spin coupling.

covalent bond a bond formed by the sharing of electrons between atoms.

cyano group the $-C \equiv N$ group.

cyanohydrin a compound with the general formula

$$R-\underset{\underset{R'}{|}}{\overset{\overset{OH}{|}}{C}}-C \equiv N$$

.

cyclization the formation of ring structures.

cycloaddition a reaction that forms a ring.

cycloalkane a ring hydrocarbon made up of carbon and hydrogen atoms joined by single bonds.

cycloalkyl alkane an alkane to which a ring structure is bonded.

cyclohydrocarbon an alkane, alkene, or alkyne formed in a ring structure rather than a straight or branched chain. The cyclohydrocarbon general formula is C_nH_{2n} (n must be a whole number of 3 or greater).

deactivating group a group that causes an aromatic ring to become less reactive toward electrophilic aromatic substitution.

debye unit (D) the unit of measure for a dipole moment. One debye equals 1.0×10^{-18} esu $\cdot$ cm. (See **dipole moment.**)

decarboxylation a reaction in which carbon dioxide is expelled from a carboxylic acid.

dehalogenation the elimination reaction in which two halogen atoms are removed from adjacent carbon atoms to form a double bond.

dehydration the elimination reaction in which water is removed from a molecule.

dehydrohalogenation the elimination reaction in which a hydrogen atom and a halogen atom are removed from a molecule to form a double bond.

delocalization the spreading of electron density or electrostatic charge across a molecule.

delocalization energy see **resonance energy.**

deprotonation the loss of a proton (hydrogen ion) from a molecule.

deshielding an effect in NMR spectroscopy that the movement of σ and π electrons within the molecule causes. Deshielding causes chemical shifts to appear at lower magnetic fields (downfield).

Diels-Alder reaction a cycloaddition reaction between a conjugated diene and an alkene that produces a 1,4-addition product.

diene an organic compound that contains two double bonds.

dienophile the alkene that adds to the diene in a Diels-Alder reaction.

dihalide a compound that contains two halogen atoms; also called a *dihaloalkane.*

diol a compound that contains two hydroxyl (—OH) groups; also called a *dihydroxy alkane.*

dipole moment a measure of the polarity of a molecule; it is the mathematical product of the charge in electrostatic units (esu) and the distance that separates the two charges in centimeters (cm). For

---

example, substituted alkynes have dipole moments caused by differences in electronegativity between the triple-bonded and single-bonded carbon atoms.

distillation the separation of components of a liquid mixture based on differences in boiling points.

double bond a multiple bond composed of one σ bond and one π bond. Rotation is not possible around a double bond. Hydrocarbons that contain one double bond are *alkenes,* and hydrocarbons with two double bonds are *dienes.*

E1 an elimination reaction mechanism in which the slow step is a self-ionization of the molecule to form a carbocation. Thus, the rate-controlling step is unimolecular.

E2 an elimination reaction mechanism in which the rate-controlling step is the simultaneous removal of a proton from the molecule by a base, resulting in the creation of a double bond. The rate controlling step is bimolecular.

electron negatively charged particles of little weight that exist in quantized probability areas around the atomic nucleus.

electron affinity the amount of energy liberated when an electron is added to an atom in the gaseous state.

electronegativity the measure of an atom's ability to attract electrons toward itself in a covalent bond. The halogen fluorine is the most electronegative element.

electronegativity scale an arbitrary scale by which the electronegativity of individual atoms can be compared.

electrophile an "electron seeker;" an atom that seeks an electron to stabilize itself.

electrophilic addition a reaction in which the addition of an electrophile to an unsaturated molecule results in the formation of a saturated molecule.

electrostatic attraction the attraction of a positive ion for a negative ion.

element of unsaturation a π bond; a multiple bond or ring in a molecule.

enantiomer a stereoisomer that cannot be superimposed on its mirror image.

enantiomorphic pair in optically active molecules with more than one stereogenic center, the two structures that are mirror images of each other are enantiomorphic pairs.

energy of reaction the difference between the total energy content of the reactants and the total energy content of the products. The greater the energy of reaction, the more stable the products.

enol an unstable compound (for example, vinyl alcohol) in which a hydroxide group is attached to a carbon bearing a carbon-carbon double bond. These compounds tautomerize to form ketones, which are more stable.

enolate ion the resonance stabilized ion formed when an aldehyde or ketone loses an α hydrogen.

epoxide a three-membered ring that contains oxygen.

ester the $-\overset{\overset{\displaystyle O}{\|}}{C}-OR$ functional group.

ether an organic compound in which an oxygen atom is bonded to carbon atoms. The general formula is $R-O-R'$. Epoxyethane, an epoxide, is a cyclic ether.

free radical an atom or group that has a single unshared electron.

free-radical chain reaction a reaction that proceeds by a free-radical intermediate in a chain mechanism—a series of self-propagating, interconnected steps. (Compare with **free-radical reaction.**)

free-radical polymerization a polymerization initiated by a free radical.

free-radical reaction a reaction in which a covalent bond is formed by the union of two radicals. (Compare with **free-radical chain reaction.**)

functional group a set of bonded atoms that displays a specific molecular structure and chemical reactivity when bonded to a carbon atom in the place of a hydrogen atom.

Grignard reagent an organometallic reagent in which magnesium metal inserts between an alkyl group and a halogen; for example, CH_3MgBr.

haloalkane an alkane that contains one or more halogen atoms; also called an alkyl halide.

halogen an electronegative, nonmetallic element in Group VII of the periodic table, including fluorine, chlorine, bromine, and iodine. Halogens are often represented in structural formulas by an "X."

halogenation a reaction in which halogen atoms are bonded to an alkene at the double bond.

halonium ion a halogen atom that bears a positive charge. This ion is highly unstable.

hemiacetal a functional group of the structure $-\overset{\displaystyle |}{\underset{\displaystyle H}{C}}\overset{\displaystyle OR}{\underset{\displaystyle OH}{\diagup}}$

hemiketal a functional group of the structure $-\overset{\displaystyle |}{\underset{\displaystyle R}{C}}\overset{\displaystyle OR}{\underset{\displaystyle OH}{\diagup}}$

hertz a measure of a wave's frequency. A hertz equals the number of waves that passes a specific point per second.

hetero atom in organic chemistry, an atom other than carbon.

heterocyclic compound a class of cyclic compounds in which one of the ring atoms is not carbon; epoxyethane, for example.

heterogenic bond formation a type of bond formed by the overlap of orbitals on adjacent atoms. One orbital of the pair donates both of the electrons to the bond.

heterolytic cleavage the fracture of a bond in such a manner that one of the atoms receives both electrons. In reactions, this asymmetrical bond rupture generates carbocation and carbanion mechanism.

homologous series a set of compounds with common compositions; for example, the alkanes, the alkenes, and the alkynes.

homologue one of a series of compounds in which each member differs from the next by a constant unit.

homolytic cleavage the fracture of a bond in such a manner that both of the atoms receive one of the bond's electrons. This symmetrical bond rupture forms free radicals; in reactions, it generates free-radical mechanisms.

Hückel's rule a rule stating that a compound with $4n + 2$ π electrons will have a closed shell electron configuration and will be aromatic.

hydration the addition of the elements of water to a molecule.

hydride shift the movement of a hydride ion, a hydrogen atom with a negative charge, to form a more inductively stabilized carbocation.

hydroboration the addition of boron hydride to a multiple bond.

hydroboration-oxidation the addition of borane (BH_3) or an alkyl borane to an alkene and its subsequent oxidation to produce the anti-Markovnikov indirect addition of water.

hydrocarbon a molecule that contains exclusively carbon and hydrogen atoms. The central bond may be a single, double, or triple covalent bond, and it forms the backbone of the molecule.

hydrogenation the addition of hydrogen to a multiple bond.

hydrohalogenation a reaction in which a hydrogen atom and a halogen atom are added to a double bond to form a saturated compound.

hydrolyze to cleave a bond via the elements of water.

inductive effect the electron donating or electron withdrawing effect that is transmitted through σ bonds. It can also be defined as the ability of an alkyl group to "push" electrons away from itself. The inductive effect gives stability to carbocations and makes tertiary carbocations the most stable.

infrared spectroscopy a type of spectroscopy that provides structural information about a molecule, based on the molecule's interaction with energy from infrared light.

initiation step the first step in the mechanism of a reaction.

initiator a material capable of being easily fragmented into free radicals, which in turn initiate a free-radical reaction.

insertion placing between two atoms.

intermediate a species that forms in one step of a multistep mechanism; intermediates are unstable and cannot be isolated.

ion a charged atom; an atom that has either lost or gained electrons.

ionic bond a bond formed by the transfer of electrons between atoms, resulting in the formation of ions of opposite charge. The electrostatic attraction between these ions is the ionic bond.

ionization energy the energy needed to remove an electron from an atom.

isolated double bond a double bond that is more than one single bond away from another double bond in a diene.

isomers compounds that have the same molecular formula but different structural formulas.

IUPAC nomenclature a systematic method for naming molecules based on a series of rules developed by the International Union of Pure and Applied Chemistry. IUPAC nomenclature is not the only system in use, but it is the most common.

Kekulé structure the structure for benzene in which there are three alternating double and single bonds in a six-membered ring of carbon atoms.

ketal the product formed by the reaction of a ketone with an alcohol. The general structure of a ketal is:

$$R-\underset{\underset{R}{|}}{C}\overset{\displaystyle OR'}{\underset{\displaystyle OR'}{\big<}}$$

keto-enol tautomerization the process by which an enol equilibrates with its corresponding aldehyde or ketone.

ketone a compound in which an oxygen atom is bonded via a double bond to a carbon atom, which is itself bonded to two more carbon atoms.

kinetics the study of reaction rates.

leaving group the negatively charged group that departs from a molecule, which is undergoing a nucleophilic substitution reaction.

Lewis theory of acids and bases a Lewis acid is a compound capable of accepting an electron pair, and a Lewis base is capable of donating an electron pair.

linear the shape of a molecule with *sp* hybrid orbitals; an alkyne.

Markovnikov rule states that the positive part of a reagent (a hydrogen atom, for example) adds to the carbon of the double bond that already has more hydrogen atoms attached to it. The negative part adds to the other carbon of the double bond. Such an arrangement leads to the formation of the more stable carbocation over other less-stable intermediates.

mass number the total number of protons and neutrons in an atom.

mechanism the series of steps that reactants go through during their conversion into products.

methylene group a —CH$_2$ group.

molecular orbital an orbital formed by the linear combination of two atomic orbitals.

molecule a covalently bonded collection of atoms that has no electrostatic charge.

multiple bond a double or triple bond; multiple bonds involve the atomic *p* orbitals in side-to-side overlap, preventing rotation.

neutralization the reaction of an acid and a base. The products of an acid and base reaction are a salt and water.

neutron an uncharged particle in the atomic nucleus that has the same weight as a proton. Additional neutrons do not change an element but convert it to one of its isotopic forms.

node a region of zero electron density in an orbital; a point of zero amplitude in a wave.

nonbenzenoid aromatic ring an aromatic ring system that does not contain a benzene ring.

nonbonding electrons valence electrons that are not used for covalent bond formation.

nonterminal alkyne an alkyne in which the triple bond is located somewhere other than the 1 position.

nuclear magnetic resonance spectroscopy a method for measuring how much energy odd-numbered nuclei absorb in the radio frequency range when the atom is exposed to strong magnetic fields. This type of spectroscopy gives information on the environment surrounding the specific nucleus.

nucleofuge see **leaving group.**

nucleophile a species that is capable of donating a pair of electrons to a nucleus.

nucleophilic substitution a reaction in which a group on a carbon atom, which has a full or partial positive charge, is displaced by a nucleophile.

nucleus the central core of an atom; the location of the protons and neutrons.

optical activity the ability of some chemicals to rotate plane-polarized light.

orbit an area around an atomic nucleus where there is a high probability of finding an electron; also called a shell. An orbit is divided into orbitals, or subshells.

orbital an area in an orbit where there is a high probability of finding an electron; a subshell. All of the orbitals in an orbit have the same principal and angular quantum numbers.

outer-shell electron see **valence electrons.**

overlap region the region in space where atomic or molecular orbitals overlap, creating an area of high-electron density.

oxidation the loss of electrons by an atom in a covalent bond. In organic reactions, this occurs when a compound accepts additional oxygen atoms.

oxonium ion a positively charged oxygen atom.

ozonide a compound formed by the addition of ozone to a double bond.

ozonolysis the cleavage of double and triple bonds by ozone, O_3.

paired spin the spinning in opposite directions of the two electrons in a bonding orbital.

parent name the root name of a molecule according to the IUPAC nomenclature rules; for example, hexane is the parent name in *trans*-1,2-dibromocyclohexane.

peroxide a compound that contains an oxygen-oxygen single covalent bond.

peroxyacid an acid of general form

$$R-\overset{\overset{\displaystyle O}{\|}}{C}-O-OH$$

π (pi) bond a bond formed by the side-to-side overlap of atomic p orbitals. A π bond is weaker than a σ bond because of poor orbital overlap caused by nuclear repulsion. Unsaturated molecules are created by π bonds.

π complex an intermediate formed when a cation is attracted to the high electron density of a π bond.

π molecular orbital a molecular orbital created by the side-to-side overlap of atomic p orbitals.

polar covalent bond a bond in which the shared electrons are not equally available in the overlap region, leading to the formation of partially positive and partially negative ends on the molecule.

polarity the asymmetrical distribution of electrons in a molecule, leading to positive and negative ends on the molecule.

precursor the substance from which another compound is formed.

preparation a reaction in which a desired chemical is produced; for example, the dehydration of an alcohol is a preparation for an alkene.

primary carbocation a carbocation to which one alkyl group is bonded.

primary (1°) carbon a carbon atom that is attached to one other carbon atom.

product the substance that forms when reactants combine in a reaction.

propagation step the step in a free radical reaction in which both a product and energy are produced. The energy keeps the reaction going.

protecting group a group that is formed on a molecule by the reaction of a reagent with a substituent on the molecule. The resulting group is less sensitive to further reaction than the original group, but it must be able to be easily reconverted to the original group.

proton a positively charged particle in the nucleus of an atom.

protonation the addition of a proton (a hydrogen ion) to a molecule.

pure covalent bond a bond in which the shared electrons are equally available to both bonded atoms.

pyrolysis the application of high temperatures to a compound.

racemate another name for **racemic mixture.**

racemic mixture a 1:1 mixture of enantiomers.

rate-determining step the step in a reaction's mechanism that requires the highest activation energy and is therefore the slowest.

rate of reaction the speed with which a reaction proceeds.

reactant a starting material.

reaction energy the difference between the energy of the reactants and that of the products.

reagent the chemicals that ordinarily produce reaction products.

rearrangement reaction a reaction that causes the skeletal structure of the reactant to undergo change in converting to the product.

reduction the gaining of electrons by an atom or molecule. In organic compounds, a reduction is an increase in the number of hydrogen atoms in a molecule.

resonance the process by which a substituent either removes electrons from or gives electrons to a π bond in a molecule; a delocalization of electrical charge in a molecule.

resonance energy the difference in energy between the calculated energy content of a resonance structure and the actual energy content of the hybrid structure.

resonance hybrid the actual structure of a molecule that shows resonance. A resonance hybrid possesses the characteristics of all possible drawn structures (and consequently cannot be drawn). It is lower in energy than any structure that can be drawn for the molecule and thus more stable than any of them.

resonance structures various intermediate structures of one molecule that differ from each other only in the positions of their electrons. None of the drawn resonance structures is correct, and the best representation is a hybrid of all the drawn structures.

R group see **alkyl group.**

ring structure a molecule in which the end atoms have bonded, forming a ring rather than a straight chain.

rotation the ability of carbon atoms attached by single bonds to freely turn, which gives the molecule an infinite number of conformations.

saturated compound a compound containing all single bonds.

saturation the condition of a molecule containing the most atoms possible; a molecule made up of single bonds.

secondary carbocation a carbocation to which two alkyl groups are bonded.

secondary (2°) carbon a carbon atom that is directly attached to two other carbon atoms.

separation technique a process by which products are isolated from each other and from impurities.

shielding an effect, in NMR spectroscopy, caused by the movement of σ and π electrons within the molecule. Shielding causes chemical shifts to appear at higher magnetic fields (upfield).

σ (sigma) antibonding molecular orbital a σ molecular orbital in which one or more of the electrons are less stable than when localized in the isolated atomic orbitals from which the molecular orbital was formed.

σ (sigma) bond a bond formed by the linear combination of orbitals in such a way that the maximum electron density is along a line joining the two nuclei of the atoms.

σ (sigma) bonding molecular orbital a σ molecular orbital in which the electrons are more stable than when they are localized in the isolated atomic orbitals from which the molecular orbital was formed.

skeletal structure the carbon backbone of a molecule.

S_N1 a substitution reaction mechanism in which the slow step is a self ionization of a molecule to form a carbocation. Thus, the rate controlling step is unimolecular.

S_N2 a substitution reaction mechanism in which the rate controlling step is a simultaneous attack by a nucleophile and a departure of a leaving group from a molecule. Thus, the rate controlling step is bimolecular.

***sp* hybrid orbital** a molecular orbital created by the combination of wave functions of an *s* and a *p* orbital.

***sp²* hybrid orbital** a molecular orbital created by the combination of wave functions of an *s* and two *p* orbitals.

***sp³* hybrid orbital** a molecular orbital created by the combination of wave functions of an *s* and three *p* orbitals.

spin-spin splitting the splitting of NMR signals caused by the coupling of nuclear spins on neighboring nonequivalent hydrogens.

steric hindrance the ability of bulky groups on carbon atoms to prevent or restrict a reagent from reaching a reaction site.

straight-chain alkane a saturated hydrocarbon that has no carbon-containing side chains.

structural isomer also known as a *constitutional isomer,* structural isomers have the same molecular formula but different bonding arrangements among their atoms. For example, C_4H_{10} can be butane or 2-methylpropane, and C_4H_8 can be 1-butene or 2-butene.

subatomic particles a component of an atom; either a proton, neutron, or electron.

substituent group any atom or group that replaces a hydrogen atom on a hydrocarbon.

substitution the replacement of an atom or group bonded to a carbon atom with a second atom or group.

substitution reaction a reaction in which one group replaces another on a molecule.

tautomers structural isomers that easily interconvert.

terminal alkyne an alkyne whose triple bond is located between the first and second carbon atoms of the chain.

terminal carbon the carbon atom on the end a carbon chain.

termination step the step in a reaction mechanism that ends the reaction, often a reaction between two free radicals.

tertiary carbocation a carbocation to which three alkyl groups are bonded.

tertiary (3°) carbon a carbon atom that is directly attached to three other carbon atoms.

tetrahaloalkane an alkane that contains four halogen atoms on the carbon chain. The halogen atoms can be located on vicinal or nonvicinal carbon atoms.

thermodynamically controlled reaction a reaction in which conditions permit two or more products to form. The products are in an equilibrium condition, allowing the more stable product to predominate.

tosyl group a p-toluenesulfonate group:

$$CH_3 - \bigcirc - \overset{\overset{\displaystyle O}{\|}}{\underset{\underset{\displaystyle O}{\|}}{S}}{}^-$$

trigonal planar the shape of a molecule with an *sp²* hybrid orbital. In this arrangement, the σ bonds are located in a single plane separated by 60° angles.

triple bond a multiple bond composed of one σ bond and two π bonds. Rotation is not possible around a triple bond. Hydrocarbons that contain triple bonds are called alkynes.

ultraviolet spectroscopy a spectroscopy that measures how much energy a molecule absorbs in the ultraviolet region of the spectrum.

unsaturated compound a compound that contains one or more multiple bonds; for example, alkenes and alkynes.

unsaturation refers to a molecule containing less than the maximum number of single bonds possible because of the presence of multiple bonds.

valence electrons the outermost electrons of an atom. The valence electrons of the carbon atom occupy the $2s$, $2p_x$, and $2p_y$ orbitals, for example.

valence shell the outermost electron orbit.

vinyl alcohol $CH_2{=}CH{-}OH$.

vinyl group the $CH_2{=}CH{-}$ group.

Wurtz reaction the coupling of two alkyl halide molecules to form an alkane.

X group "X" is often used as the abbreviation for a halogen substituent in the structural formula of an organic molecule.

ylide a neutral molecule in which two oppositely charged atoms are directly bonded to each other.

Zaitsev's rule states that the major product in the formation of alkenes by elimination reactions will be the more highly substituted alkene, or the alkene with more substituents on the carbon atoms of the double bond.

IA								
1 H Hydrogen 1.00797	IIA							
3 Li Lithium 6.939	4 Be Beryllium 9.0122							
11 Na Sodium 22.9898	12 Mg Magnesium 24.312							
19 K Potassium 39.102	20 Ca Calcium 40.08	21 Sc Scandium 44.956	22 Ti Titanium 47.90	23 V Vanadium 50.942	24 Cr Chromium 51.996	25 Mn Manganese 54.9380	26 Fe Iron 55.847	27 Co Cobalt 58.9332
37 Rb Rubidium 85.47	38 Sr Strontium 87.62	39 Y Yttrium 88.905	40 Zr Zirconium 91.22	41 Nb Niobium 92.906	42 Mo Molybdenum 95.94	43 Tc Technetium (99)	44 Ru Ruthenium 101.07	45 Rh Rhodium 102.905
55 Cs Cesium 132.905	56 Ba Barium 137.34	57 La Lanthanum 138.91	72 Hf Hafnium 179.49	73 Ta Tantalum 180.948	74 W Tungsten 183.85	75 Re Rhenium 186.2	76 Os Osmium 190.2	77 Ir Iridium 192.2
87 Fr Francium (223)	88 Ra Radium (226)	89 Ac Actinium (227)						

	58 Ce Cerium 140.12	59 Pr Praseodymium 140.907	60 Nd Neodymium 144.24	61 Pm Promethium (145)	62 Sm Samarium 150.35	63 Eu Europium 151.96
Lanthanide Series						
Actinide Series	90 Th Thorium 232.038	91 Pa Protactinium (231)	92 U Uranium 238.03	93 Np Neptunium (237)	94 Pu Plutonium (242)	95 Am Americium (243)

Atomic weights shown below the symbols are based on Carbon–12. The numbers above the symbols are the atomic numbers.

							VIIIA
							2 He Helium 4.0026
		IIIA	IVA	VA	VIA	VIIA	
		5 B Boron 10.811	6 C Carbon 12.01115	7 N Nitrogen 14.0067	8 O Oxygen 15.9994	9 F Flourine 18.9984	10 Ne Neon 20.183
		13 Al Aluminum 26.9815	14 Si Silicon 28.086	15 P Phosphorus 30.9738	16 S Sulfur 32.064	17 Cl Chlorine 35.453	18 Ar Argon 39.948
28 Ni Nickel 58.71	29 Cu Copper 63.546	30 Zn Zinc 65.37	31 Ga Gallium 69.72	32 Ge Germanium 72.59	33 As Arsenic 74.9216	34 Se Selenium 78.96	35 Br Bromine 79.904	36 Kr Krypton 83.80
46 Pd Palladium 106.4	47 Ag Silver 107.868	48 Cd Cadmium 112.40	49 In Indium 114.82	50 Sn Tin 118.69	51 Sb Antimony 121.75	52 Te Tellurium 127.60	53 I Iodine 126.9044	54 Xe Xenon 131.30
78 Pt Platinum 195.09	79 Au Gold 196.967	80 Hg Mercury 200.59	81 Tl Thallium 204.37	82 Pb Lead 207.19	83 Bi Bismuth 208.980	84 Po Polonium (210)	85 At Astatine (210)	86 Rn Radon (222)

64 Gd Gadolinium 157.25	65 Tb Terbium 158.924	66 Dy Dysprosium 162.50	67 Ho Holmium 164.930	68 Er Erbium 167.26	69 Tm Thulium 168.934	70 Yb Ytterbium 173.04	71 Lu Lutetium 174.97
96 Cm Curium (247)	97 Bk Berkelium (247)	98 Cf Californium (251)	99 Es Einsteinium (254)	100 Fm Fermium (257)	101 Md Mendelevium (258)	102 No Nobelium (259)	103 Lr Lawrencium (260)

H	
2.1	

Li	Be
1.0	1.5
Na	Mg
0.9	1.2
K	Ca
0.8	1.0
Rb	Sr
0.8	1.0
Cs	Ba
0.7	0.9

B	C	N	O	F
2.0	2.5	3.0	3.5	4.0
Al	Si	P	S	Cl
1.5	1.8	2.1	2.5	3.0
Ga	Ge	As	Se	Br
1.6	1.8	2.0	2.4	2.8
				I
				2.4

Many elements have been omitted to emphasize the basic pattern of electronegativity variation.